BUILDING SOIL

A DOWN-TO-EARTH APPROACH

NATURAL SOLUTIONS *for* BETTER GARDENS *and* YARDS

ELIZABETH MURPHY

COOL
SPRINGS
PRESS
Home and Garden Experts™
MINNEAPOLIS, MINNESOTA

Quarto is the authority on a wide range of topics.

Quarto educates, entertains and enriches the lives of our readers—enthusiasts and lovers of hands-on living.

www.quartoknows.com

First published in 2015 by Cool Springs Press, an imprint of Quarto Publishing Group USA Inc., 400 First Avenue North, Suite 400, Minneapolis, MN 55401 USA

Telephone: (612) 344-8100 Fax: (612) 344-8692

Cool Springs Press titles are also available at discounts in bulk quantity for industrial or sales-promotional use. For details write to Special Sales Manager at Quarto Publishing Group USA Inc., 400 First Avenue North, Suite 400, Minneapolis, MN 55401 USA.

ISBN: 978-1-59186-619-0

Library of Congress Cataloging-in-Publication Data

Murphy, Elizabeth, 1976– author.
  Building soil : a down-to-earth approach : natural solutions for better gardens & yards / Elizabeth Murphy.
    pages   cm
  ISBN 978-1-59186-619-0 (sc)
  1. Soils. 2. Soil management. I. Title.
  S591.M984 2015
  631.4—dc23
                                                                              2014036112

Acquisitions Editor: Mark Johanson
Project Manager: Madeleine Vasaly
Art Director: Cindy Samargia Laun
Cover Designer: Erin Seaward-Hiatt
Book Designer: Wendy Holdman
Photographer: Crystal Liepa Photography (except as noted on page 200)
Illustrator: Bill Kersey

Printed in China

10 9 8 7 6 5 4

▪ CONTENTS ▪

INTRODUCTION

Why soil? For me, it's a love affair. I find nothing more satisfying than digging between my garden rows and pulling up a handful of loose, dark, and sweet-smelling dirt. It is a handful that breaks apart with perfect spongy softness, a bounty of red wrigglers falling between my fingers. In that scoop of living earth is immediate gratification that I must be doing something right in my garden, though exactly what and how are sometimes a mystery. That mystery of the soil, for me, is part of what makes gardening an art.

If you are reading this book, your love affair may have already begun. There is something in our very nature that draws us to soil. It is the ground on which we walk and creates the contours and colors of where we live. In France, it is *terroir*, which defines the qualities of a wine. The world over, it is earth itself. In our lives, we are grounded when our roots stretch deep in a community or place. From it, we pull the very food that sustains and nourishes us. It is the mysterious medium in which a seed is planted and new life begins, to bloom, bear fruit, or grow tall.

Nonetheless, how does a love for soil help to translate fertilizer labels, determine when and how much to water, or help build compost? What can we do with the materials we have at hand to build the soil in a sustainable way? Why would we test our soil and what does it mean? Most of all, how can we create more verdant, productive, and beautiful gardens by working less and enjoying the process more?

In the age of industrial agriculture, our intuitive knowledge of how to build healthy soils as the foundation for healthy gardens needs rediscovery. In my own search to rediscover soils, I often wanted a practical guide to the what, when, and how of caring for my soil. This book aims to do just that, combining what I've learned over a decade of farming, gardening, soils research, and teaching into a clear and practical guide for building your garden from the ground up. Far from being a complicated process, this comes down to a few basic principles that can guide almost any gardening decision.

My love of soils began with my first vegetable patch in our Oakland, California, neighborhood community garden. In the middle of inner-city concrete, hilling up rows of dirt and turning messy compost satisfied an unfulfilled need. We packed the worm bins with scraps from our kitchen and saw our plots transformed with the gooey black compost the worms provided. Building up the soil in a once-vacant lot brought existing fig trees and lilac bushes back to life. Soon, the eighth-acre garden started to attract birds, butterflies, and raccoons. One day, I was laid flat on the ground by thousands of bees swarming just overhead, darkening the sky with an inescapable hum; as one of the few green islands in our neighborhood, the garden had naturally attracted the drifting bees. The garden began to serve as a focal point for the neighborhood as well. People wandered through

My neighborhood community garden started my love affair with soil.

it on evening walks or rested in the shade on a sunny day. Transforming this spot in the city by transforming the soil seemed nothing less than miraculous to me.

This experience led me to work on numerous small, organic farms throughout the Pacific Northwest, learning valuable lessons from both farms that focused on caring for the soil and farms that focused on the crop. Though both could pull in a good harvest, there was a difference in how the farm looked and felt, whether there was bare earth compacted by tractor tires or bountiful hedgerows lining the fields. I wondered how we could best match caring for the soil with ensuring a good crop while meeting the farmer's bottom line and not breaking our backs in the process.

Eventually, on a winding search for answers to these questions, I found myself in a soils laboratory at the University of California, Davis. There, I joined the frontlines of soil organic matter research, investigating the material that is the foundation of healthy soils. We strained the limits of technology to get a closer look at soil organic matter. There were a lot of questions that begged for answers. What is its exact chemical composition? How does it interact with both the living and the nonliving parts of the soil? And why exactly does it have such a profound effect on everything from feeding the world to pollution to climate change?

When I left the lab to become an agricultural extension agent in southwestern Oregon, soil organic matter remained a recurring theme in my new job as well. Answers for many of the questions that flooded my desk often came back to building healthier soils. Adding soil organic matter to create a more resilient garden and farm helped with everything from weed problems to chicken influenza. For farmers, the key to a successful business lay in reducing the costs of inputs, such as fertilizer, fuel, and water. Caring for the soil first, by building organic matter, reduced the need for inputs, and in the long term could determine the profit-breaking point for organic and conventional farms alike. For Master Gardeners, learning to care for the soil meant that, over time, labor could be reduced and enjoyment increased.

In classes and consultations, I found that people were hungry to understand how to care for the part of their garden that was underground. As gardeners, we instinctively know that soil nourishes our plants. We can sense the difference between a living, vibrant, dark black fertile soil and dried-out, nutrition-starved dirt. All too often, though, we don't have a clear guide on how to put this intuitive knowledge into practice. The legacy of simplified fertilizer percentages has left people confused about soils and the part of their gardens that they can't see. Although the fertilizer bag can play an important role in building and nurturing soils, in reality, soil is much more than a composite of chemicals. It is a living, breathing part of your garden that has the same needs as any other living thing. Treating your soil like a living system actually makes gardening much easier. By providing the things any living organism needs—namely, food, shelter, water, and air—the soil starts to take care of itself. In the process, it takes care of your gardens, lawns, and trees.

Whether I'm teaching or gardening, I try to demystify the world of soils by developing clear guidelines for how to mimic and rely on natural processes. In doing this, we can speed up changes to the soil that may take decades if left to nature into a few short growing seasons. My home garden, which started as a half-acre of weeds and heavy clay, has given me a living laboratory for the how-to techniques and tools presented in this book. A busy schedule and a lot of ground meant I needed to find the fastest, easiest, and cheapest way to build my soil. Feeding the soil first meant using what was most easily at hand, from cardboard to compost.

The world over, people are finding that this holistic view of feeding the soil translates into feeding their families and their communities. At the same time, feeding the soil means recycling waste into garden bounty. In this process, you create sustainable systems in your own backyard. Since sustainable soils filter water, require less chemicals, and capture carbon from the atmosphere, building living soils in our backyards and our communities is a small but important part of addressing global environmental problems, from water pollution to climate change.

Scientific discoveries underpin our current understanding of how and why building living soils improves our gardens and our world. Knowing

Understanding that your soil is a living system is the first step in becoming a soil-based gardener.

the details of soil science, however, is not necessary to learn how to nurture the living soil. For generations, our ancestors did just this, using traditional and sustainable gardening and farming techniques. For this reason, I've structured this book for the practical gardener, focusing on the fundamental soil-building tools needed to care for more productive, and less time-consuming, gardens, lawns, and trees.

The first chapter starts with an explanation of the living soil, focusing on how we recognize a healthy soil and what we can and can't change through our practices. The second chapter provides a set of clear principles for gardening from the ground up. You can use these as guidelines in making almost any gardening decision. The chapters that follow detail how these principles are applied to all your gardening activities, such as fertilizing, tilling, weed management, and irrigation. Finally, we put it all together in Chapter 7 by outlining the month-by-month tasks for the soil-based gardener. In this chapter, I'll explain exactly what to do if you want to start a garden from scratch or improve an existing garden.

Building soil takes some initial sweat equity, but once the living soil is balanced, the system practically runs itself.

Ultimately, the information in this book translates to soil-based gardening methods that won't break your back, are good for the environment, and create high-yielding and beautiful gardens of all shapes and sizes. Because the tools and processes explained in the following chapters are based on fundamental principles, they can apply to almost any garden situation. They are used to create classic landscape gardens, grow a high-yielding orchard, nurture naturally beautiful lawns, raise your household veggies, or run a profitable farm. Hopefully, reading these pages will transform you as the gardener as well, opening new eyes to see not just the plants but also the living system that grows them.

CHAPTER 1

A SOIL GROWER'S FOUNDATION

I'm looking out over my backyard, iced tea in hand. It is a late summer day, and the vegetable garden is a riot of color and texture. Hollyhocks and cosmos have grown to giant proportions and reach skyward between the large leaves of winter squashes. Tomatoes and beans twine upward on trellises. Carrots, beets, and cucumbers are hidden treasures between the broad green leaves of collards and kale. Clovers cover pathways between the beds. The herb garden is a jumble of lavender, rosemary, thyme, and calendula, buzzing noisily with the bees that scavenge the plants' nectar. Small fruits swell on apple trees that are spread between clumps of currants and raspberries. The only thing that I can't see when I look out over my garden is the soil.

Nonetheless, beneath this verdant garden jungle is a world equally, if not more, alive. The living soil is a beating, breathing, eating, growing, digesting, dying, and throbbing organism. More accurately, it is a collection of organisms so abundant that they rival all of earth's aboveground creatures in number and type. One teaspoon of this microscopic world contains vastly more individuals than the human populations of New York, London, Hong Kong, Tokyo, Mexico City, Chicago, and Moscow put together. This living system functions at once as the lungs, filter, and food source for the planet. Almost every molecule of earth's water, air, and nutrients has passed through the soil at some point in its cycle. In doing so, these materials are consumed and transformed by the billions and billions of organisms that inhabit the top 12 inches of gardens, orchards, and lawns. Although invisible to us, the teeming life below ground creates the conditions necessary for thriving life above ground. Know this and you will know the secret to successful gardening.

For me, the complexity of a soil can be boiled down to a basic premise: grow healthy soil to grow healthy plants. An afficionado of garden science, I've researched soil in minute detail, from its chemical composition to the way clay minerals interact. In my garden, however, I only use two basic soil concepts to grow healthy soils: *What does the living soil need?* and *What does the nonliving soil provide?* This chapter spells out the needs of the living soil, while giving you the tools to know a healthy soil when you see it. In my own experience, once I gained the eyes to see soil health, taking care of the soil became second nature.

A garden thrives when built from the ground up. Diverse, vertical layers of living plants cover the living soil, which is the foundation and source for whole-garden bounty.

IT'S ALIVE!

So what is soil anyway? We walk on it every day. We avoid it when it turns to mud. We move it around to plant seeds or weed garden beds.

In any soil-science class, the first things a student learns are the four fundamental components of a soil. If you looked at your soil in profile, you would see that almost half, by volume, is composed of minerals. These are the little bits of broken-up rock that took centuries of wind, flowing water, ice, and rain to be worn down to particles that range from small to really, really small. You can think of this as the nonliving backbone of the soil.

Looking at a soil profile, we easily see the solid minerals that make up about half of the soil. The other half—empty pores filled with water and air—is harder to see. Though barely visible (the living and organic portion of the soil is less than 5 percent by volume), it drives almost all of the soil's, and thus our garden's, processes and functions.

Almost the entire remaining half of the soil is composed of the area between the minerals, collectively termed *pore space*. The size of this pore space also ranges in size, from large channels we can see to small microscopic spaces, and it is filled with a mixture of water and air. The amount of water or air that fills the pore space fluctuates depending on rainfall or irrigation but is roughly equal parts of the two. Water clings to the sides of the solid soil, while air fills the spaces in between.

These three components—minerals, water, and air—make up nearly all of what we see when we look at our soil. With only these three ingredients, however, we would be living on a barren planet. It is the remaining 5 to 10 percent of the soil that actually makes plant life possible. This fraction, the fourth component of soil, is soil organic matter.

Soil organic matter is all the material in the soil that currently is or once was living. This includes the leaves that fall in the autumn and the manure tilled under in the spring. It includes active plant roots and the unrecognizable, decayed roots from five years ago.

What's in Soil?

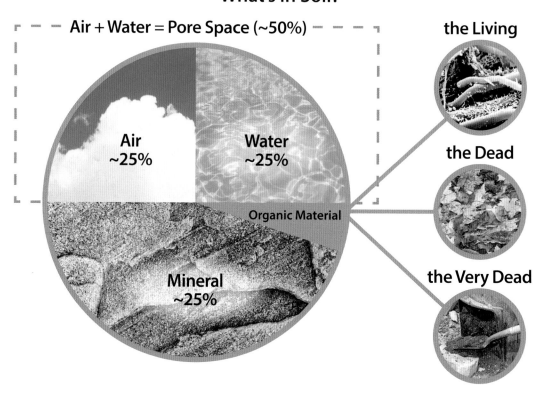

Most of a soil is made up of minerals, and the water and air-containing pore space between them. A small but mighty portion (less than 5 percent) of the soil is organic matter, made up of living, dead, and very dead plants and organisms. Organic matter gives soil its dark black color and is the basis for building soil health.

It includes the living and dead bodies of microorganisms and the material that they excrete. It includes the worms, spiders, and nematodes that we can and can't see. It even includes the cardboard and paper that we use to mulch the soil surface. Yes, that cardboard was once a tree and that tree was once living.

Despite its small percentage, the organic part of the soil exerts control over almost all of a soil's properties. Organic matter changes what you can do with a difficult clay or sandy soil. It increases how much water the soil can hold and when it can be released to plants. It even stores and recycles plant nutrients in the bodies of living and dead bacteria and fungus. Without organic matter, a soil is merely broken-down rock.

Within the relatively small fraction of soil that is organic matter, only 5 percent is actually alive. In terms of whole-soil compositions, the means that less than 0.5 percent of the entire soil is living! The remainder of organic matter in the soil is in some stage of decomposition. Together, the living and decomposing organic portions of your soil are a tangle of live, newly dead, and long-dead organisms and plants that hold the soil system together. Nonetheless, if such a tiny fraction of the soil is actually alive, why do we call it a *living soil*?

This mighty 0.5 percent of a soil that lives and breathes at any given time is a powerhouse of activity whose effect vastly outweighs its proportional size or weight. It's the living engine of the garden that feeds plants by constantly consuming and excreting nutrients. It creates a rich ecosystem that protects the garden from pest and disease. A living soil is even responsible for the rich, dark, earthy smell of soil—a smell that modern medicine has associated with antidepressant properties. Our fundamental job as gardeners is to work with, instead of against, the living soil. We do our best not to harm it while, at the same time, try to improve the conditions it needs to thrive. Like most life on earth, the living soil has four basic requirements: air, water, food, and shelter. By understanding how the nonliving soil meets these needs, we can create the perfect conditions for a living soil to thrive.

THE MINERAL SOIL

Let's start by taking a look at the solid part of the soil: the minerals, made up of broken-down bits of rock. The character of the mineral soil depends on a variety of factors, such as which rock it came from and how long it was beaten by wind, rain, or waves. Whatever the origin or process, the end result is small particles that you may or may not be able to see without a microscope.

When you sift out the larger pieces of rock and gravel in your soil, the small particles left behind are named based on three categories of particle size: sand, silt, and clay. Sand is the largest particle size; we can easily see its gritty shape with our naked eye. Silt and

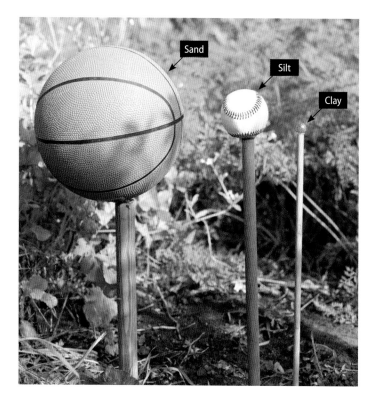

Though we can't see it with our naked eye, size differences between soil particles are immense. Sand (pictured here as a basketball) is a thousand times bigger than clay (pictured here as a marble). Silt (baseball) is of intermediate size.

clay are too small to see individually. Silt is small, and clay is really, really small. Although we can't observe silts and clays directly, we can feel the difference between them with our fingers. Silt has a smooth, silky feel when wet. Clay sticks to your fingers and can be molded into shapes.

It may not seem like these size categories would make much difference, but in fact the amounts of sand, silt, and clay in your garden soil determine almost all of its properties. These are the qualities of a soil that you may be able to modify a bit through your management but that, like our own inherited traits of height or eye color, we are more or less stuck with. Because of this, as a starting point for any garden, it is important to know which particles make up the mineral fraction of your soil.

The amount of sand, silt, and clay in a soil is known as soil texture. Soil texture is so important that it is actually used to name soil types. Using the USDA soil textural triangle, the relative percentages of sand, silt, and clay give your soil a name, such as loam, sandy loam, or clay. These names directly describe how much of each particle size is in a soil, but they also are useful for describing the properties of your soil and talking shop with other soil aficionados.

A quick way to determine soil texture is with the do-it-yourself soil jar test (see pages 18 and 19). The USDA online soil survey website (www.websoilsurvey.sc.egov.usda.gov), though not always site specific, can tell you how your soil is mapped. The best soil texture results come with a laboratory soil test, as described in Chapter 3.

Why does texture matter so much? For one thing, it is one of the only properties of a soil that we can't change. If you have a clay soil, you are stuck with a clay soil. Period. I'm often asked whether we can truck in sand to fix a clay soil or vice versa. The answer, quite simply, is "No!" If you do this, nine times out of ten, you will end up with concrete, not soil. Why? Sandy soil has the perfect surface for clay to stick. When this happens, concrete, instead of soil, is made. Trucking in a load of sand simply can't mimic the natural soil-building process.

Differences in soil texture also determine a few important decisions that you will tailor to your garden. These differences can be boiled down to the differences in pore space (how much water and air a soil can hold), chemical charge (how easily nutrients can stick to the soil), and soil strength (how easily a soil is compacted). In the end, this helps you fine-tune how much and when to water, fertilize, and till your soil.

Soils that have too much sand, silt, or clay create gardening challenges. If you've ever tried growing vegetables in beach sand, you might know what I am talking about. Sand does not easily hold onto water, nutrients, or organic matter. Anything added to a sandy soil drains straight down in a process called leaching. Silt can easily clog up pore spaces, creating poorly drained, waterlogged soils. Clay sticks to itself and anything else. It holds onto a lot of water and nutrients, but its attraction is so strong that plants might not be able to use them. Clays are also shaped like flat books that are easily compressed when wet. When this happens, we get hard-to-work, compacted soil.

Sometimes, we lovingly describe a good soil as loamy, a name based on soil texture. The reason that loamy soils are so sought after by gardeners is that they are the perfect combination of sand, silt, and clay. When working together, sand will provide large, connected channels for drainage, while clays will hold onto water and nutrients. Sand will provide resilience to compaction, while clays will create good, crumbly soil structure.

We can choose many things about how we garden, but we can't choose the texture of the soil. Because of this, it's important to get to know a particular patch of ground before making big decisions. Although you can't change the texture of your garden soil, nurturing the living soil, with techniques in this book, can overcome many of the limitations of soil texture to grow lush gardens. A key part of this process is sometimes matching what we choose to grow to what our soil is best suited for.

My garden has several distinct areas with very different properties. One area, near the fence line, is low lying and clay rich. To complicate things, it is periodically inundated by the neighbor's irrigation. It took me a few tries to observe that this area could only support plants adapted to harsh flood and drought conditions. Realizing this, I planted daylilies and wild currants, which covered the previously barren soil with a food- and

continued on page 21

WHAT'S YOUR TEXTURE?

A quick-and-dirty way to get a handle on the texture of your garden soil is with the simple jar test. This test lets you estimate the percentages of sand, silt, and clay in your soil by watching how these particles settle out of water.

Imagine washing your dog in the bathtub. The gritty beach sand in her fur always falls to the bottom first, but the water will stay dirty for a long time because of the clays that remain floating. This is because the large, heavier sand particles settle out quickly, while the smaller, almost weightless clays stay in suspension for a long time. The jar test uses the same idea to measure soil texture.

For this test, you'll need a clear glass quart jar, a timer, water, a marking pen, and a shovel to take your soil samples. Take a soil sample by scraping recognizable plant material and fluffy organic matter off the soil surface until you reach mineral soil (usually less than ½ inch deep). Dig about 8 inches down into the soil and scrape a trowel along the edge of the hole to sample the entire depth. To get a good representation of your soil, take several samples throughout your garden and mix them in a bucket. Take a handful of the soil you mixed in your bucket and fill your glass jar about halfway full. Mark a line at the level of your dry soil. Now fill the jar about two-thirds full of water and shake it vigorously for three minutes. At the end of three minutes, set the jar down and start your timer. At thirty seconds, mark a line on your jar where the soil has settled out. This is a rough estimate of the amount of sand in your sample. Continue timing the jar, and three minutes after you made the first mark, make

Name that soil! Use the US Department of Agriculture soil textural triangle to determine the name of your soil. Each side of the triangle represents percentages of sand, silt, or clay. Choose two of these and follow the lines up from their relative percentages. Where the lines cross is the name of your soil.

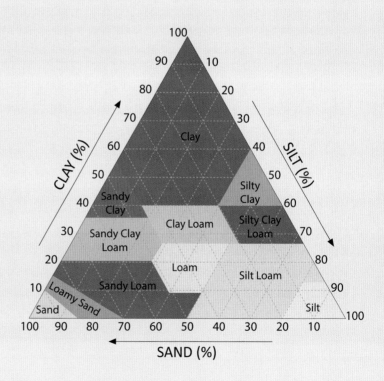

a second one. This is the amount of silt in your soil. Look at the original line that you made, and use this to estimate what percentage of sand and silt make up the total amount of the sample. You can assume that the remaining amount not accounted for in the sand and silt portions is clay.

Now take these three percentages and look at the USDA textural triangle to determine your soil texture. To do this, start with the percentage of sand you estimated for your soil. Look along the bottom edge of the triangle and find the line that corresponds to this percentage. Follow the sand line up and to the left. Now look at the right-side edge of the triangle. Find the line that corresponds to the percentage of silt that you estimated for your soil. Follow the silt line down and to the left. The point where these two lines intersect determines the name of your soil. For instance, if you have 40 percent sand and 40 percent silt in your soil, these lines cross in the "loam" region. Therefore, you have a loam soil. Trace the horizontal line that crosses at this point back to the left-hand edge of the triangle. This represents the percent clay in your soil. Doing this for the above example, we can determine that our loam soil also has 20 percent clay.

Are you lucky enough to have a loamy soil? Remember that this test is just a quick way to get familiar with your soil texture. Did your jar test tell you that you have more sand or more clay? These general questions will guide you in understanding how to fine-tune your soil-based gardening approach.

How to Perform a Jar Test

2. Mark the level of your dry sample on the jar. Fill the jar two-thirds full with water and shake vigorously for three minutes.

1. To take a soil sample, scrape away the top organic layer until reaching mineral soil. Dig an 8-inch-deep hole and scrape along its depth to sample about a cup of soil. Take several samples throughout the garden and blend in a bucket, then fill a glass jar about half full.

3. Let the soil settle for 30 seconds and then mark a line at the top of the soil. Wait three minutes then mark the top line again.

4. After marking the second line on your jar, you should have two clear layers: sand (lower) and silt (upper). Estimate the percentage of each particle in relation to the original level of soil in the jar. The particles above the silt line are clays. As this size of particle takes a long time to fully settle, estimate the percentage of clay by difference after accounting for sand and silt.

A GARDENER'S GUIDE TO THE MINERAL SOIL

The amount of sand, silt, and clay in a soil determines how it behaves. Knowing this, we can best use water, fertilizers, and tillage to build healthy soils. Extremes of sand or clay can create gardening nightmares, while loamy soils are the stuff of green garden dreams.

Property	Sandy Soil	Loamy Soil	Clay Soil
Drainage and wetness	Excessively well-drained. It's like pouring water through a sieve. Water and nutrients move too fast for plants to use. You don't have a problem with wetness, but your plants may be getting parched.	Just right—not too wet, not too dry.	Slow to drain. Your ground can easily become waterlogged or have wetness problems in low-lying areas.
Compaction	You can drive a semi-truck over your garden without much effect.	It's good to be cautious with wet soils, but this soil can take a bit of punishment.	Be extremely careful when your soil is wet. Drive a tractor over your soil now, and your grandchildren will see the tire ruts. If you dig in your garden after a rain, you're doing more harm than good.
Nutrients and fertility	Nutrients are not stored in this soil. You must provide the nutrients your plants need, as they need them. Tends to be acidic.	Good nutrient holding ability. Good turnover of organic matter to release new nutrients.	Full of nutrients and organic matter. Wetness can cause problems with availability. Tends to be alkaline.
Water for plants	Does not hold water. Your plants use water as it comes and then wait for the next rain or irrigation.	Holds the maximum amount of water available for plants.	Holds the most water, but because it is in small pores, it may not be available to plants.
Soil structure	Not much structure.	Good crumbly, soil aggregates. This structure supports organic material and the living soil.	Clays stick together to make structure, but this structure is vulnerable to compaction. Clays may stick together in blocks or plates, which cause drainage problems.
Temperature	Warms up fast in spring for an early crop.	Warms up reasonably fast in the spring.	Takes longer to warm up in spring. You have to wait to plant that early spring crop because of germination or wetness problems.

One of the keys to successful gardening is matching your gardening goals with the soil's potential. Carrots grow deep and sweet in a sandy soil, while brassicas, such as cabbage and broccoli, can handle the cooler and wetter conditions of heavy clay.

continued from page 17

flower-producing oasis. At the other end of the property, the ground is slightly raised. For some reason, this area is much sandier than anywhere else in my garden. Because of the lighter soil, this area provides perfect conditions for drainage-sensitive garden gems, such as asparagus and blueberries.

SOIL STRUCTURE

Soil structure is the way that the individual sand, silt, and clay particles stick together. Good structure provides homes for soil organisms and pockets that hold on to soil water and food. Poorly structured soil, on the other hand, may impede drainage, restrict root growth, lack fertility, or not have enough air and water for living organisms.

One way to understand soil structure is to think of how the soil breaks apart in our hands. What we want feels similar to a broken-up bit of brownie. It melts in our hands—soft, crumbly, and delicious. These crumbs of soil are known as aggregates and are glued together by sticky clays, organic matter, plant roots, bacterial slime, and fungal threads. Gardeners often talk about a *friable* soil. This satisfying word describes the feeling of an easy-to-dig, crumbly soil, a soil that is chock full of these stable, small aggregates.

Unlike with soil texture, we have a great deal of control over soil structure. By enhancing soil structure, many of the limitations of soil texture are overcome. For instance, when small clay particles form larger aggregates, soils drain better.

Improving soil structure increases the availability of food, shelter, water, and air—all of the basic needs of the living soil. Because of this, better soil structure means the soil supports more living organisms. Soil organisms, at the same time, build soil aggregates.

We can feel good soil structure with our hands. It is soft and crumbly like a brownie. These crumbs, called aggregates, make the best homes for soil organisms.

Improving soil structure, largely through adding soil organic matter, starts a win-win feedback that keeps improving the soil condition over time. This is why building soils through organic matter is the basis of any soil-growing program.

THE SOIL ECOSYSTEM

When we look at our gardens as a whole—aboveground, belowground, living, and non-living—we are looking at the garden ecosystem. This whole system view can include aspects of our homes, our lives, and the society that our garden touches. An ecosystem perspective is the foundation of the sustainable-design principles of permaculture, a philosophy that follows natural cycles, rather than trying to change them.

In many ways, the soil is the foundation of the garden ecosystem. It controls the flow of nutrients and water to garden plants. This happens primarily through the activity of soil organisms. The soil ecosystem is full of many types of organisms, which vary in size, food preference, job skills, and housing requirements. They range from single-celled bacteria to microscopic protists to recognizable worms and insects.

Diverse soil organisms provide a diversity of functions. Living soil organisms release nutrients to plants by shredding, grazing, and digesting organic matter. Nitrogen-fixing rhizobium bacteria that live on the roots of bean-family plants literally pull nitrogen out of thin air into the soil. Fungi that live in beneficial relationships with plant roots, known as mycorrhizal fungi, increase the length of a plant root to improve delivery of water

THE SCOOP ON WORM POOP

The earthworm's ability to transform soil is almost incomprehensible. These powerful creatures eat their way through an enormous amount of earth, ingesting more than their weight in soil (up to two to thirty times more!) daily. Depending on how active they are, this adds up to between 100 to 2,000 pounds of soil per 100 square feet of your garden that passes through an earthworm gut in a single year. Worm movement through the soil creates burrows for plant roots and water to pass. In burrowing, worms also mix organic matter from the surface, throughout the soil.

Earthworms excrete ingested soil as perfect round, crumbly stable aggregates—the ideal soil structure for which we, as gardeners, strive. These aggregates, known as earthworm casts, bring with them all the benefits of a friable soil structure, such as improved drainage and a greater ability to hold onto water and organic matter. An added benefit is that these casts are power-packed packages of nutrients that greatly increase soil fertility. Earthworms are so efficient at mixing the soil and improving its structure that they take over the work of rototiller, tractor, and fork in no-till gardens.

Feeding the soil with organic matter invites earthworms into our garden soils. Encouraging these industrious soil animals replaces the time and labor spent digging a hard soil year after year. In less time, and with less work on your part, these soil engineers will help you create the fluffy, friable soil of your dreams.

and phosphorus. Together, a healthy community of soil organisms keeps pests and plant diseases in check.

This diversity of organisms has a diversity of needs. You wouldn't feed a human baby and your pet goldfish the same bananas. An orca whale and a monarch butterfly don't like to live in the same places. The more complex we make our gardens, the more likely we are to meet this diversity of needs.

In no way does this idea of complexity mean complicated. In fact, complex gardens are actually simpler to manage. Complex gardens let weeds get out of hand sometimes. Complex gardens mix wildflowers with vegetables. Complex gardens may let part of last year's crop rot on the soil surface, while mixing well-aged compost into a carefully sculpted seed bed. Complex lawns may be okay with a dandelion or two.

Bacteria nodules

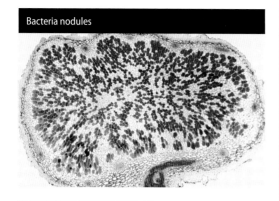

Spider, *Walckenaera acuminata*

Beneficial nematodes

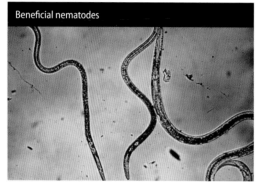

Fungus-eating protozoa

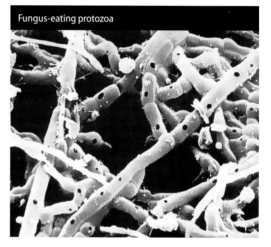

Basidiomycete fungi

The soil ecosystem is teeming with microscopic life, such as bacteria, nematodes, fungi, protists, and spiders. These organisms control the flow of resources to plants and drive the productivity of the whole-garden ecosystem.

By meeting this diversity of needs, the living soil becomes more diverse. By becoming more diverse, the functions of the soil ecosystem are replicated in a variety of different ways by a variety of different organisms. In a thriving city, there is any number of skilled car mechanics from which to choose. If one is booked, then there is another you can call on in case of an emergency. You may find one whose specialty is brake jobs and one makes art out of rare, import cars. The same principles of redundancy and diversity hold true for a healthy soil ecosystem.

Caring for the soil ecosystem as a whole means meeting the four basic needs of any living organism. If we broadly think about meeting the living soil's need for food, water, air, and shelter, then the ecosystem begins to more or less take care of itself. As long as we keep things interesting, as long as our garden and lawns have assorted shapes, contours, and colors, then we should be supplying the necessary diversity for diverse soil ecosystems. Throughout the rest of this book, the guidelines for soil-based gardening will come back again and again to how to meet these four basic requirements for living organisms.

Understanding these four foundational needs of your living soil provides the template every gardener needs for nurturing the living, breathing component of the soil. Although different soil textures require slightly different approaches and cautions, the basic needs of food, water, air, and shelter in a living soil can either be met or improved with the addition of organic matter.

RECOGNIZING HEALTHY SOIL

To create healthy soils, we first need to recognize a healthy soil. We often can't see the organisms that drive our garden's growth, but we can observe their effects with our five senses. These observations are called soil-quality indicators. Observing these indicators over time gives us a way to measure how we are doing at building soil health. More importantly, these observations start to give us the eyes to see soil health from an ecosystem perspective.

A word commonly used describe soil health is *soil tilth*. Tilth refers to a combination of desirable soil properties, which converge to create that good garden soil that we can feel, smell, and almost taste. Good tilth is a fluffy handful of friable, dark brown earth, breaking apart into round crumbs. Good tilth is the satisfaction of your gardening fork sliding easily into the bed of carrots. Good tilth is the spongy perfection of a slightly moist soil.

The way you assess soil quality and tilth in your garden depends on your goals. At a basic level, keeping these indicators in mind gives you a way to recognize the health of your soil. In this way, you can recognize a good quality soil when you see it. In my own experience, I have found that using these guidelines helps me to put my gardening and soil-building goals into perspective.

At a more detailed level, the gardener who loves keeping records can actually measure changes in soil health over time. To compare measurements over time you need to take them at the same time of year and under roughly the same conditions. If you want to get really scientific, be sure to take several measurements each time in different, randomly chosen spots. Average the measurements together to get a truer reading of the health of your soil across the garden. Use the *Soil Quality Report Card* in this chapter to record your observations and keep track of how your garden changes as soil health improves.

These indicators are meant as common-sense tools to help you notice differences. They are a guide to help you learn to listen to your living soil. As such, you are free to change, interpret, or discard them in the best way that suits your needs. In fact, soil-quality indicators change from state to state and region to region (see the Additional Resources section for more about soil quality and specific guidelines that may exist in your region). You may even think of some new and original quality indicators for your own site, such as, how often does that ditch by the fence fill with water; how many cat prints do clean up after a summer rain; how many passes does it take to rake a smooth seedbed. Let these questions open your eyes and your mind to the living soil. Have fun with them!

Soil Drainage

Soil drainage, the ability of water to percolate through your soil, is also known as infiltration. Observing soil drainage is easy; just look at the soil surface. Next time you have a hard rain, put your raincoat and go into your garden. Is water pooling on the surface? Is it running off in rills and gullies and carrying soil with it? Instead, is your garden soaking up the rain like a sponge?

After the rain, check your garden again. How long does water remain ponded on the surface? When the soil dries out, do you notice cracks or crusts on the surface? In most cases, soils with good structure should drain twenty-four hours after a big rain. Cracking and crusting on the soil surface is also an indicator of poor drainage.

When water does accumulate at the soil surface, you can also watch whether your garden is built to prevent erosion. You've worked hard to build healthy living topsoil and you don't want it to be carried away by a spring storm into streams or sewage drains.

Cracks or crusts on the soil surface indicate that water ponded due to slow soil drainage.

MY SOIL QUALITY REPORT CARD

Use this worksheet to keep track of soil quality indicators and soil health over time.

Date:	Garden location:
Time:	Plants grown:
Soil moisture: dry/moist/wet	General notes:

Indicator	Rating (1–10)	Poor	Fair	Good
Soil drainage		Water ponded on soil surface more than 24 hours after rain. Cracks and crusts on soil surface.	Some ponding after rain event.	Water absorbed by soil. No ponding.
Soil moisture		Soil is moist: _____ days after the last rain/irrigation.		
Soil aggregates		Soil breaks into flakes, clods, or powdery dust, or the soil has no structure at all.	Some aggregate crumbs are visible.	Soil full of aggregate crumbs that are stable under pressure.
Soil compaction		Wire flag won't penetrate the soil. Bends readily.	Wire flag meets some resistance. Difficult to push through the soil.	Wire flag penetrates soil easily to a depth of _____ .
Plant growth		Plants are stunted, yellowish, or delayed in development. Plants don't look uniform across the garden.	Some plants are stunted, yellowish, or delayed.	Plants are vigorous, green, and uniformly healthy across the garden.
Plant roots		Roots not deep or well developed. Some roots may seem rotten or self-pruned.	Some branching structure and fine roots present. Moderately deep rooting depth.	Fully developed, branching, and deep root structure. Many fine, white roots present.
Soil biology		1. _____# organisms_____. # types in _____ minutes. 2. _____# earthworm burrows/casts. _____# of earthworms.		
Plant residues		If residue is present, it is fibrous and recognizable. May have a sour smell. No dark staining of fingers.	Some residue present and some evidence of decomposing residue. Some dark brown color to stain.	Residues and residue decomposition apparent. Sweet smell to soil. Dark stain on fingers.

Adapted from the Willamette Valley Soil Quality Score Card, EM 8711 (Oregon State University Extension. 1998).

Mulch and plantings on the soil surface help to slow down runoff water and keep it from carrying your soil away. Similarly, orient garden beds perpendicular to slopes to slow and catch water, allowing it to infiltrate and recharge your soil.

Soil Moisture

Closely related to soil drainage, soil moisture is the amount of water held by your soil. Too much and your soils are waterlogged. Too little and your soils are stressed by drought. The ideal moisture condition feels like a wrung-out sponge. The amount of moisture your soil can hold over time is important in determining its ability to provide water to plants and microbes. As with soil drainage, your soil texture has a built-in ability to hold on to soil water. Organic matter, however, functions exactly like a sponge and can greatly increase how much water your soil can hold.

As you build your soil with organic matter, it should hold more water for longer periods of time. An easy way to evaluate this soil-quality indicator is to simply note how long the water lasts after a good saturating rain or irrigation. Touch the soil to feel for the ideal spongy, damp soil-moisture condition. Notice how long it takes before your plants show signs of water stress. Can water from a deep rain or irrigation last two days, a week, or longer?

Soil Aggregates

The importance of good soil aggregation can't be stressed enough. Improvements to soil aggregation go hand in hand with improving the living component of your soil. Soil aggregates that indicate a living system are round and crumbly. Good soil aggregates are also stable under pressure. Like most of the following indicators, assess soil aggregates under good soil moisture conditions, when soils are not too dry or too wet.

To determine the quality of your soil aggregates, take a shovelful from the top 6 to 12 inches of your soil. Using your fingers, gently break the soil apart and see what sort of pieces you end up with. Does the soil break apart in clods or powdery flakes? Instead, does it break into rounded crumbs? These are the soil aggregates. Note what percentage of the shovelful you took breaks apart into these crumb-like structures.

Now, take some of these aggregates and press them between your fingers. How easily do they break apart under pressure? Use a bit of water to gently wet the aggregates, and press them again. Do they still hold up under pressure?

The best-quality soil breaks apart into rounded aggregates instead of large blocks, powdery flakes, or single grains. The best aggregates will hold their shape under moderate pressure when moist and when wet. With more pressure, they should break apart nicely. Sensitize your hands to what a well-aggregated soil feels like and you can quickly assess soil condition at any time. Building your soils should lead to a soil with increasingly more of this friable crumb structure every year.

Soil Compaction

Soil compaction describes exactly what it sounds like: a dense, hard, compact soil. In the case of our gardens this is not desirable. Remember that soil is made up of about half minerals and half pore space. That pore space is vital for supplying air and water to roots and organisms. It is also extremely important for the passage of water, plant roots, and critters through the soil profile. A compacted soil is one in which the minerals have been pushed together, and the space for air and water is squeezed out. Since the soil is a living ecosystem that depends on water and air, compacting soil is like squeezing out the life.

To assess changes in soil compaction, use a 12- to 18-inch wire flag from the hardware store. Push the flag into the soil and note how far it goes and how much pressure it takes. In a severely compacted soil, the flag will bend before it gets far. In a soil with moderate or deeper soil compaction, the flag will meet some resistance before fully penetrating. In a deep, noncompacted soil, the wire flag will go straight down into deeper soil layers without much resistance at all.

Soil Compaction Test

Use the wire flag test to determine differences in soil depth and compaction in different spots of the garden. When compacted, the wire flag will bend as it meets hard soil layers.

Plant Growth

It may seem obvious, but how well your plants are growing is one of the best indicators of your soil's health. The vigor of plant growth indicates the soil's ability to provide nutrients, water, and a suitable growing medium. Observing your plants and your garden is also one of the best tools you have to understand reasons for good or bad yields, pest problems, or optimum microsites for specific plants. It's also fun.

Observe plant health and vigor at the height of the growing season, before flowering. First, look across your garden and lawn and notice the uniformity of growth. Are there areas that look worse or better than others? There may be an underlying soil-quality issue driving these imbalances. Assess the color of the plant. How green are the leaves? Greener leaves indicate healthier plants. Are the plants growing faster or slower than other years under similar conditions? Do you observe pests or diseases? These problems often occur because plants are susceptible to them due to poor soil conditions.

Plant Roots

Plants' roots are great indicators of soil compaction or drainage problems. When roots are not free to grow it is usually because they cannot penetrate compacted layers or waterlogged soils. Roots are the

The health and vigor of a plant's rooting system is a great indicator of the health and vigor of the soil. Healthy soils grow plants, like this goldenrod, with root systems that are deep, large, and branching.

mechanisms by which a plant gets nutrition and water from the soil. They are also hot spots for soil organisms. Shedding dead roots feed the soil with organic matter, while live roots leak yummy sugars and carbohydrates to the surrounding soil. For these reasons, the roots of your garden plants tell you a great deal about the health of your soil. A functioning soil is filled with living roots.

Choose a garden plant or even a well-placed weed to observe plant root growth. Use your shovel and fork to dig wide around the plant, trying to pull up as much of the intact roots as possible. When looking at your plant roots, observe how deep the root system extends. Are roots branching and complex, with many fine roots? Are new roots healthy and white? What we want to see is a complex root structure, fully branching and developed. A healthy soil grows plant roots that extend deep below the topsoil and are covered with new, white roots, too numerous to count. In fact, if you have trouble digging up the entire root system because it extends too deeply, that is a great sign.

Soil Biology

Much of the living soil is invisible to us without a microscope, yet there are some creatures, like earthworms and spiders, that we can see. Since the soil ecosystem is all connected, the living soil that we can see gives us an indication about the vastly greater number of living organisms that we can't see. In this way, counting the number and type of spiders, ants, termites, and worms seen in a shovelful of soil indicates how alive our soil is.

Soil biology is best assessed when soil organisms are the most active, when soils are not too hot, too cold, or too dry. Moist soils in mid-spring or mid-autumn are ideal for this measurement. Dig a shovelful of soil down to 6 inches. Drop the soil on a material with a white background (a white sheet or a piece of poster board works well). Break it apart and inspect it for a given amount of time. Count the number and types of organisms you see in the given time span, ideally two to four minutes. Ants, termites, and spiders move fast, so you have to be quick to catch them. When counting organism types, it is not necessary to know their names, just to note the many types you see. Generally, a soil is in good condition if you see three or more individuals of each of three or more types of organisms in the space of two to four minutes.

Another way to observe soil biology is to focus exclusively on earthworms, since they play a crucial role in building soil health. First, get down on your hands and knees at the soil surface to look for evidence of earthworm engineering, detected by the casts and burrows that they leave behind. Earthworm casts are the round aggregates created as the worms digest and excrete organic matter. Burrows are small holes at the soil

Earthworms active at the soil surface leave behind small, round piles of soil, known as casts, as well as visible burrows. Count these at the soil surface to evaluate soil biological activity.

surface. After counting the casts and burrows on the surface, dig down to 6 inches and count the number of earthworms you find in your shovelful of soil. A rich, living soil will have ample casts and burrows at the surface and more than three to five worms in a shovelful.

Plant Residues

Decaying plant residues in the soil are also an indicator of soil biological activity, although an indirect one. Our primary aim as gardeners is to add to and build organic materials in the soil. Counterintuitively, though, we don't want to see this material accumulate in the soil, at least not in the form we added. A lot of recognizable plant residues means that organic matter is not decaying fast enough.

A living soil will quickly transform recognizable organic materials, such as manure, plant roots, or leaf mold, into dark, spongy soil organic matter. In the soil, organic materials are shredded into bite-sized pieces by the ants and spiders, mixed into the soil by worms, and bathed in chemical baths by fungi and bacteria. In this way, organic matter is broken down into food energy to pump the living soil engine. Because of this, the condition of residues in our soil gives a good indication of our living soil's activity.

Observe plant residues at least a month after incorporating a cover crop, residues, or other organic material. This is best done in moist summer or early fall soils. Dig a small trench, about 6 inches deep, at the surface of your soil. Break apart the soil and look for residues. Put your nose in there, too, and take a good whiff. Moisten a bit of the soil in your hand, rub it between your fingers, and see if it leaves a dark stain.

A good quality soil will have plant residues in various stages of decomposition, some pieces you can recognize and some pieces you can't recognize. You want the soil to leave a dark stain on your fingers when wetted. This is the organic matter that has broken down into bits too small to see individually. A soil that is properly decomposing will have a sweet, earthy smell. This is the smell of bacteria at work.

A soil that is not efficiently decomposing residues will have abundant, recognizable fibrous plant materials. The soil may have a sour smell, like a wet compost pile that you haven't turned in a long time. These undecomposed plant residues can be an indication of a waterlogged soil or a soil that is too dry, both conditions that limit soil organism activity.

THE FINAL TOUCH

The key to growing gardens from the ground up is getting down to ground level. To know what's going on in your soil, touch it, smell it, see it, and feel it. Unlock the secret to growing healthy soil by recognizing a living soil. At first, it is hard to imagine that the ground beneath our feet is filled with microscopic life, but over time, it is possible to actually feel when the living soil is active and alive. This foundation, combined with the principles in the next chapter, begins the lifelong journey of nurturing our gardens by growing healthy, living soils.

CHAPTER 2

GUIDING PRINCIPLES FOR THE SOIL GROWER

THE SECRETS OF SOIL ORGANIC MATTER

Now that we have the tools to recognize healthy soils, our next step is to learn how to create healthy soils. When I first started gardening, I was overwhelmed by the details. I had trouble keeping track of what, when, where, and why and with making sense of all the different information I was reading.

Over time, I figured out that the secret to successful gardening was to simplify. Instead of focusing on the details, I zoomed out to view my garden as part of a whole ecosystem. In this way, I could focus on nurturing principles that applied for almost every aspect of my garden and home.

Looking at soil health from a holistic ecosystem perspective comes back to the basic idea of providing food, shelter, air, and water for the living soil. With this template, we can ask, *What does the soil have and what does it need?* The best way to provide these essential needs for the living soil is often to do nothing at all, letting one part of the whole ecosystem provide for another.

For instance, if I let leaves fall from my trees into the garden bed, they provide shelter (mulch) and food (organic matter) for soil organisms. As the soil transforms the leaves, it releases nutrients back to the tree and builds long-lasting organic matter in the soil. The organic matter holds onto more water that the tree can use. It also supports a greater amount of mycorrhizal fungi, which increases the length of tree roots and delivers nutrients from deeper in the soil profile.

Working with a whole system perspective, each of us will find what works in our own garden through trial and error. Failures are often the checks and balances of an overall productive system. Following principles that foster living systems simplifies our day-to-day gardening activities down to a few simple guiding principles for building soil health.

IT COMES DOWN TO SUSTAINABILITY

The word *sustainability* means different things to different people. Can your kids support themselves after college? If yes, they're self-sustainable. Does your current lifestyle have longevity? If yes, it's sustainable. More and more, sustainability refers to environmentally friendly practices. Does a practice contribute to the longevity of earth's natural resources? If yes, then it's sustainable. In soil, sustainability also refers to how well this living system takes care of itself. Can our soils, and thus our gardens, remain fertile and productive without much help from us? Then they're sustainable.

Sustainability is best pictured by looking at what goes into and what goes out of our gardens. Sustainable gardens and soils minimize what leaves the system, thereby

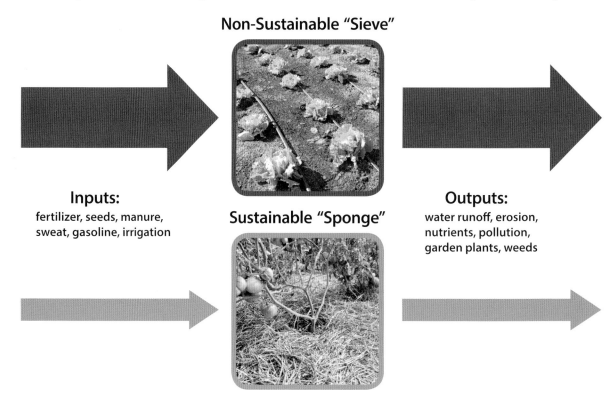

Non-Sustainable "Sieve"

Inputs:
fertilizer, seeds, manure, sweat, gasoline, irrigation

Sustainable "Sponge"

Outputs:
water runoff, erosion, nutrients, pollution, garden plants, weeds

Sustainable systems minimize what enters and leaves the system, relying on internal processes to recycle and retain resources. Non-sustainable systems don't hold onto resources and require continual inputs from the outside.

minimizing what needs to be added. This boils down to whether your garden is a sponge or a sieve. Does your garden hold onto water, nutrients, and other resources so that you add less and less over time, like a sponge? Or do these inputs simply pass through the garden like a sieve, continually requiring you to add more and more to meet plant demand?

As a sponge, sustainable gardens also maximize the capture of natural resources, such as rainwater, sunlight, or nitrogen, from the atmosphere. Many permaculture principles for natural farming are based on this concept. Building swales (contoured trenches) across the gentle slope of a garden keeps water on the landscape. Keyhole beds are shaped to maximize the amount of sunlight available to plants. Harvesting rainwater from rooftops increases the capture area of our garden sponge and brings our homes into the whole garden ecosystem.

Our goal in building sustainable soils is to continually strive to make our gardens into sponges, rather than sieves. To accomplish this, we promote natural processes as much as possible. In nature, living soils are adapted to hold onto water and nutrients, the precious currency of life. A healthy living soil will recycle and store these materials, keeping them in the soil system for plants to use and for gardens to thrive.

Permaculture techniques, such as this keyhole bed that bends toward the sun, maximizes the capture of natural resources by the whole-garden ecosystem.

Sustainable systems, like this old-growth forest, are self-perpetuating sponges. They are built to capture and hold onto sunlight and water. Nutrients are continually recycled between the active soil ecosystem belowground and the productive forest aboveground, maintaining a productive, layered, and diverse forest ecosystem.

Building healthy living soils is the key to building sustainable gardens. In doing so, we reduce the inputs that we need, whether labor, fertilizer, or irrigation. This not only saves us time and money but also conserves natural resources to create a more sustainable world.

ORGANIC MATTER: THE SOIL SPONGE

Amazingly, there's a single solution that provides food, shelter, water, and air for the living soil. The miracle cure is organic matter. Organic matter is anything living or anything dead that was once living. If it has the gooey, sticky signature of life, then its organic matter.

Despite its relatively boring name, the effect of organic matter in the soil is anything but boring. Organic matter is, essentially, the sponge of the soil. It captures and holds onto resources like water and nutrients. It changes soil structure and builds friable, loose soils. It provides basic food to plants and living organisms. From an organism's point

Organic Matter Provides . . . . . . for the Living Soil

Water

Shelter

Food

Air

Organic matter gives fertile soil its rich, black color. It's the secret ingredient that makes soils into resource-retaining sponges, providing the living soil with food, shelter, water, and air.

of view, organic matter turns the soil environment from a disaster-relief shelter into a luxury condominium, leading to bigger, healthier, and more productive populations of living soil organisms.

Organic Matter Is Food

First and foremost, I like to think of organic matter as food for the soil. When stranded on a deserted island, a person who is fed has the energy to make his or her own shelter, explore the island, and build signals to summon rescue help. If living organisms are fed, then they also have the energy to improve their own homes by improving soil condition.

Organic matter provides the raw materials for building the bodies and tissues of organisms and plants. Decomposition, the rotting of organic matter, is an important process in your soil, your compost pile, and sometimes your refrigerator. It starts when tiny microorganisms munch on raw organic matter and excrete it as something new. This process of transformation releases important nutrients, such as phosphorus and potassium. It also releases a large amount of energy that fuels the living activity of the soil engine. Your plants and soils need this nutrition and energy to grow and thrive. In this way, when you feed the soil, you also literally feed your plants.

Organic Matter Holds Water

In addition to being food for the soil, organic matter stores soil water. Like a sponge, it has a huge amount of pore space and a lot of surface area. It is able to absorb vast amounts of water to release when conditions become dry. Even a small amount of organic matter in the soil can greatly increase the amount of water your soil can hold after rain or irrigation to provide more water for microbes and plants in the living soil. This is particularly helpful for sandy soils with little water-holding ability.

Organic Matter Lets the Soil Breathe

Take a handful of mineral soil and a handful of well-rotted compost. Which is lighter? The properties of organic matter that make it a good sponge are the same properties that make it fluffy and full of air. All that space in between the little bits and pieces of soil organic matter, when they are not filled with water, are filled with air. This gives the soil more space to breathe, providing oxygen for living organisms and plant roots. Adding fluffy organic matter to the soil, drainage and other problems of compaction are also improved. Since waterlogged soil limits the oxygen available to living organisms, improving drainage is an essential part of maintaining a well-aerated soil.

ORGANIC MATTER AND GREENHOUSE GASES

By weight, organic matter in soils is roughly half carbon. Carbon is one of the main ingredients of greenhouse gases responsible for climate change. Building soil health, we increase the amount of organic matter stored in the soil. This means that more carbon is stored in the soil. Across the globe, this amounts to a huge amount of carbon. Because of the delicate balance of earth's resources, more carbon in the soil means less carbon in the atmosphere. Less carbon in the atmosphere means reduced greenhouse gases.

Storing carbon in soil by increasing organic matter is known as *carbon sequestration*. It is an important part of the solution to reduce the effects of climate change. In this way, adopting sustainable practices for our garden soils affects sustainability of the planet.

Organic Matter Builds Better Homes

Like us, the living creatures in your soil need homes. They need shelter from predators and cover from hard rains. They create cozy spaces that have the right temperature, moisture, and oxygen level for their taste. Organic matter is also what homes are made of in the living soil. As critters munch on organic matter, they create the glue that binds soil particles together. These bound particles, which we know as aggregates, are apartment complexes for the many and varied organisms in the soil.

Mulching the soil surface with organic matter also shelters surface-dwelling organisms. An organic cover on the soil protects it from the impact of exposure to hard rains, baking sun, or freezing weather. In this way, precious topsoil is protected from the elements and from being carried away by wind or water.

SOIL-ORGANIC-MATTER GOGGLES

Because organic matter is the key ingredient for meeting the living soil's needs, I base almost all of my gardening decisions on this fundamental soil component. I think of this perspective as having soil-organic-matter goggles, looking at the effects of everything I do in terms of whether the soil gains or loses organic matter. In this way, I can build whole-garden sustainability by viewing conditions from the living soil's perspective.

Often, we think of soil organic matter as something we add to a soil. The revolutionary change to this perspective is realizing that organic matter is also something that we keep, promote, and grow through our practices. Amending with manure increases organic matter, but so does interplanting lettuces, growing a green manure, or reducing how often we till.

We build organic matter levels in the soil by adding raw organic materials as amendments. We also build organic matter by letting nature take its course, keeping fallen leaves and plant residues in the garden. Finally, we build organic matter by maintaining what is there and preventing activities that remove it.

Organic matter is lost when we remove plant material from the soil (grass clippings, leaves, garden harvest, weeds). It is also lost if wind or water erodes the topsoil. With decomposition, the process that makes the living-soil engine hum, a portion of organic matter is lost to the atmosphere. For this reason, tillage, which promotes rapid organic matter decomposition, is one of the ways we lose soil organic matter.

ORGANIC MATTER: MIRACLE CURE?

Unfortunately, organic matter alone will not fix all of a garden's problems. Moreover, too much organic matter, especially when added as an amendment, can be too much of a good thing. For instance, adding 6 inches of manure, instead of 1 to 2, will overwhelm the living soil's ability to decompose it. Too many rich nutrients can encourage pests and diseases. Manures, in particular, can cause problems with salt buildup.

Even with organic matter, other limiting soil conditions, like drainage, compaction, and fertility can keep the living soil from thriving. One time, I accidentally created a water-logged section of my garden by digging sunken paths in a clay soil. I treated this section like the rest of my garden, tilling back in the previous year's garden residue and adding heaps of compost. The next season, little of what I had added had decomposed. The soil stayed cold, wet, and mucky with half-rotten squash and tomato vines. The undecomposed organic matter was a haven for aphids over the winter.

The following season, though the garden bed was relatively dry in the droughty summer, my tomatoes and peppers were slow, stunted, and struggling. Because of waterlogged conditions, the soil organisms had not used the organic matter that I had so thoughtfully added. It wasn't until I drained this section, by digging swales across the contour of my yard, that this part of my garden started functioning. High and dry, the soil organisms were able to convert the mucky organic residues into deep black earth. This process gradually improved the damage from water-logged conditions. In this way, only once I addressed the underlying limitation of soil wetness could my garden take advantage of the powerful benefits of organic matter.

Since organic matter is continually lost, it needs continual replenishment. In nature, organic matter is replenished when plants grow and return to the soil as residues and litter. In our gardens, organic matter is replenished when we add amendments, grow and till in a green manure, or return plant residues to garden beds. Chapter 4 will delve into the specific do-it-yourself ways to replenish organic matter in this way.

Knowing whether your actions grow, add, keep, or deplete organic matter in the garden is the foundation of soil-based gardening. Use the table in this chapter as a starting guide to view common garden practices through organic-matter goggles.

SOIL ORGANIC MATTER	
To grow a healthy soil, look at garden practices in terms of whether they add to or take away from organic matter in the garden.	
ADD or KEEP	**LOSE or DEPLETE**
Less tillage	Heavy, repeated tillage
Keeping residues after harvest	Removing residues after harvest
Returning leaves and grass clippings to soil surface	Removing leaves and grass clippings
Building beds parallel with slopes	Soil erosion
Mulching soil surface	Soil crusting
Planting cover crops and green manures	Bare fallow periods
Supplying adequate fertilizers and lime	Overfertilizing
Supplying adequate water	Working a wet soil (compaction)
Planting nitrogen-fixing plants (bean family)	Weeding
Adding composts and manures	Removing garden harvest without amending soils
Cutting lawns to higher stubble heights (>1 inch)	Cutting grass lawns to the soil surface

GUIDING PRINCIPLES FOR THE SOIL GROWER

Our primary mission as soil-based gardeners is to build soil health by nurturing a vital, living soil. Understanding this broad view, and looking at our gardens, lawns, orchards, and farms as a whole sustainable system, we can distill this understanding into principles that can guide us in day-to-day decisions.

1. **Use what you have.**

   A sustainable system is self-reliant. Follow sustainability principles by using what is most readily available. Before looking to the outside to meet garden needs, see how you can use what you have at hand. Instead of irrigating more, conserve water or harvest it from rooftops. Prune and chip hedges instead of carting in wood chips. Using what you have reduces costs and labor, making it easier to improve the whole system.

Building organic matter in soils includes reducing the ways we lose it. Instead of bagging tree leaves in the fall, add them back to the garden as mulch to keep the organic matter in the soil.

2. **Go with the flow.**

   Work with, rather than against, the garden ecosystem. Watch and learn from the natural systems in your garden. Observe the garden throughout a whole year to know how it changes with the seasons. Where does water naturally flow and collect? What areas stay warmest and driest? Are there ways you can adjust your garden to work with the natural patterns and processes you observe?

Harvesting water from rooftops increases the resources captured by the garden, reducing the amount of irrigation water needed.

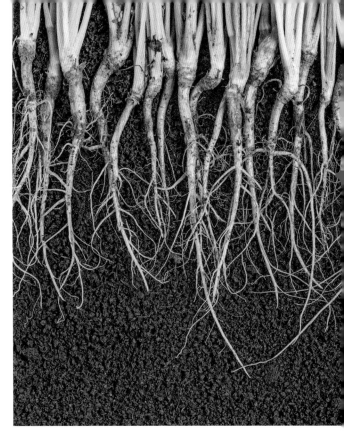

3. **Give more than you take.**

Your garden won't thrive without being replenished. Remember that inputs must balance output in a sustainable system. For all the losses of organic matter by harvest, weeding, mowing, or tilling, find a way to add organic matter back to the system. Minimize what you "take" by minimizing losses of organic matter, nutrients, water, and energy. Maximize what you "give" by capturing these materials through natural processes.

4. **As above, so below.**

If you wonder what's happening underground, take a look at what is happening aboveground. Aboveground plant health and vigor mean below-ground health and vigor. In general, roots below ground add even more organic matter to the soil than the crop aboveground.

Healthy roots are as great in mass, length, and volume as aboveground shoots. Roots add organic matter and exude juicy soil food exactly where the living soil needs it. Because of this, growing cover crops and green manures are some of the best ways to amend a soil.

5. **Make your life simple by making your garden complex.**

Diversity is the foundation of a well-functioning soil ecosystem, which self-regulates and prevents many problems before they start. To reap these benefits, meet the needs of a diverse ecosystem by building a diverse garden. A diversity of plants, rotations, additions, and gardening styles keeps life interesting and also offers varied types of food and shelter to the varied organisms in your soil.

6. **Follow the Goldilocks rule.**

Keep your soil happy by avoiding extremes. Your soil likes it "just right." Remember that too much of a good thing is not always better. Avoid over-amending, over-fertilizing, or overwatering. The living soil likes to be wet, but not too wet; warm, but not too warm; well fed, but not overfed. It prefers to have a sustained supply of food and nutrients over time, rather than living in states of feast and famine.

Diversity makes ecosystems resilient, productive, and interesting. Diverse gardens aboveground equate to diverse soil ecosystems belowground.

7. **Disturb less.**

   The living soil is healthier and happier if allowed to grow and thrive without disturbance. Tilling your soil, though not without benefits, is essentially akin to razing a whole neighborhood of houses and asking the inhabitants to rebuild anew. Bare soil can lead to losses of topsoil and soil organic matter. Pesticides, both organic and conventional, can pose the greatest disturbance for the living soil. They essentially annihilate whole populations of soil organisms, turning back the clock on efforts to build a live, healthy soil.

8. **Never till wet soil.**

   Particularly if you have clay-rich soils, the damage of working soil when it's wet far outweighs the benefits. Though the temptation in early spring may be high, avoid tilling, amending, planting, or otherwise compacting a soggy soil. Waiting for the proper moisture conditions will lead to healthier plants and earlier harvests in the long term.

9. **Keep the soil covered.**

Cover—in the form of mulches, decaying organic matter, or living plants—shelters the living organisms in your soil from harsh conditions and prevents erosion of precious topsoil. A bare, uncovered soil is also a lost opportunity for adding organic matter to the soil.

10. **Manage your landscape as a whole.**

Zoom out beyond individual garden beds, plants, or blades of grass. How can all the parts of your garden contribute to the health of the whole? A bank of flowers feeds the bees that pollinate orchard fruits and garden veggies. Hedges or trees can provide shade and shelter from wind or noise.

Disturb less by using a sheet mulch instead of a tiller to build garden soils. When we keep the soil ecosystem intact, earthworms and other soil critters can till the soil for us.

CHAPTER 3

ALPHABET SOUP

NUTRIENTS AND SOIL-NUTRIENT TESTS DEMYSTIFIED

Gardeners I meet every day are hungry to feed their soils. They instinctively know that this is the foundation for great gardens. Across the board, though, people are confused about the complexity of soils, soil tests, and fertilizers. The letters and numbers on fertilizer bags and soil test reports can have us all scratching our heads.

Luckily, the fundamentals of feeding the soil are fairly simple. If we zoom out, feeding the soil essentially comes down to providing a balanced diet of organic matter foodstuff. In an active, living soil, this, in turn, provides the essential nutrients for garden plants.

Nonetheless, it can take some time of actively building soil health before we can rely solely on organic matter for soil fertility. Soils may be severely damaged from past activities that have disturbed topsoil or depleted nutrients. It may be difficult to find high-quality organic amendments that supply the right mix of nutrients.

To build soils into sustainable systems, we often have to search out and import concentrated nutrient sources in the form of manures or ready-to-use fertilizers. When we do this, we rely on the sometimes confusing figures on soil reports and fertilizer bags to guide us. Even though those letters and numbers aren't what we actually find in a living, breathing soil, they can help us get on the right path to soil health.

This chapter will explain what this alphabet soup actually means so that we can use the information to build soils into healthy living systems. We'll do this by first looking at the

Nitrogen Deficiency

Phosphorus Deficiency

If a plant doesn't get all the essential nutrients it needs from the soil, it will show deficiency symptoms. These symptoms, such as yellowing leaves when there is not enough nitrogen or reddish-purple leaves with a lack of phosphorus, help in diagnosing soil fertility problems.

food needs of plants and the living soil. We'll discuss how and why to take a soil test to find out if the soil food needs are being met. Finally, we demystify the soil test report and, in the process, highlight the roles of important nutrients in plant and soil nutrition.

PLANT FOOD

Like us, plants need good nutrition to survive and grow. Some nutrients are essential, meaning that the plant can't live without them; if a plant lacks an essential nutrient it will show symptoms of poor health, known as deficiency symptoms. Like scurvy for sailors lacking vitamin C, these deficiency symptoms, such as stunted growth, discolored leaves, or mottles and spots, indicate that the soil is not meeting all of the plant's nutritional needs.

There are seventeen essential plant nutrients. Of these, carbon, hydrogen, and oxygen are supplied by air and water. The remaining fourteen essential plant nutrients are supplied by the soil. Gardeners are most familiar with the revered trinity of primary macronutrients: nitrogen, phosphorus, and potassium, collectively known as NPK from their chemical symbols, N (nitrogen), P (phosphorus), and K (potassium). The series of three numbers, separated by dashes, that you will find on a fertilizer bag represents the percentage by weight of each of these nutrients. Fertilizing is primarily concerned with adding NPK, since these nutrients are most often limited in soils.

Despite the emphasis on NPK, the remaining macronutrients (calcium, magnesium, and sulfur) and the eight essential micronutrients are all *essential* for plant growth and survival. A soil can provide all the NPK a plant could ever want, but if it doesn't offer the tiny bit of boron it requires, the plant simply will not grow.

Plants take up nutrients through their roots. This means that plant food must be dissolved in soil water. Quick-release and chemical fertilizers immediately dissolve into soil water, making them instantly available to plant roots. This is useful when a quick fix is needed to address a deficiency. On the other hand, it also means they are instantly vulnerable to being lost when water drains out of the soil.

Nutrients from organic amendments and slow-release fertilizers stick around longer in the soil. They are plant-available only after microbes decompose them. We rely on living systems to make these nutrients available and to hold them in the soil. Fertile soils hold nutrients in the actual bodies of living and dead organisms, in spongy organic matter, or on the surfaces of soil minerals.

PLANT NUTRITION	
The following nutrients are necessary for plant growth.	
Macronutrients	
Primary Nutrients	
Nitrogen	1.5%
Potassium	1.0%
Phosphorus	0.2%
Secondary Nutrients	
Calcium	0.5%
Magnesium	0.2%
Sulfur	0.2%
Micronutrients	
Chlorine	0.01%
Iron	0.01%
Boron	0.002%
Manganese	0.005%
Zinc	0.002%
Copper	0.0006%
Molybdenum	0.00001%
Nickel	0.000001%

What's in a plant? In addition to the carbon, oxygen, and hydrogen a plant gets from air and water, a plant is made up of the essential nutrients it gets from the soil. If the soil can't supply all of these nutrients, plant health will suffer.

SOIL FOOD

Since the soil organisms do the actual work of building healthy soil, feeding the living soil is the gardener's number-one priority. In this way, the living soil makes nutrients and water available for plants, keeps diseases in check, and improves soil tilth for root growth. When we focus on feeding the soil first, we feed plants in more sustainable and long-lasting ways, reducing the costs of supplemental fertilizers.

So what does the soil eat? Soil organisms require the same essential elements as plants. However, unlike plants, which get their carbon from the atmosphere, soil bugs need to eat carbon that is in organic matter. Since bacteria, fungi, and other soil animals are mostly made up of carbon and nitrogen, these are the nutrients most needed for the living soil to grow in size, number, and activity.

Coincidentally, carbon and nitrogen is the stuff of organic matter everywhere: plant residues, kitchen scraps, dead leaves, barnyard waste, and even our own human bodies. The quality and composition of different sources of organic matter differs, but, in some way or another, if it is organic material it becomes food for the soil ecosystem.

Feed the Soil to Feed the Plant

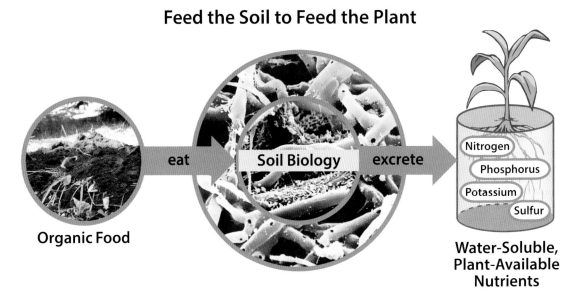

eat **Soil Biology** excrete

Organic Food

Nitrogen
Phosphorus
Potassium
Sulfur

**Water-Soluble,
Plant-Available
Nutrients**

Microbes use soil nutrients for growth. As they grow, they improve soil conditions, while storing soil nutrients in their bodies. When microbes die, they release nutrients to feed plants.

NITROGEN

When I teach my classes, I'm always preaching "the gospel of nitrogen." Can I get an amen? Although all the soil nutrients are important for great gardens, our successes and failures often come down to how we handle this single one. Making, capturing, and keeping nitrogen through soil-building activities is one of the gardener's most important jobs.

Nitrogen is often the difference between a booming garden and one that's gasping for air. It is responsible for rapid plant growth and the green color of leaves and shoots. The bodies of soil organisms are also built of nitrogen, which means that nitrogen is essential for the living soil to function. This includes the critical job breaking down organic matter to release other essential plant nutrients.

Too much nitrogen, on the other hand, is also a problem. With excess nitrogen, plants grow too fast for their own good, falling over on leggy stems. They put extra energy into leaf and shoot production and not enough into fruit and flowers. The quality of vegetables and fruits also suffers. A carrot grown in nitrogen-loaded soil loses its sweetness and tastes pretty bad. Too much nitrogen fertilizer (yes, this includes organic fertilizer) is also

Nitrogen makes gardens green, bountiful, and productive. Supplying adequate nitrogen, however, can be a balancing act. It is the key ingredient for luscious, verdant gardens, but too much can cause pest, quality, and environmental problems.

a common culprit in water pollution, making streams, rivers, and oceans inhospitable to fish.

Nonetheless, we focus on nitrogen for one simple reason: nitrogen is on the move. Unlike most of the other nutrients that cycle in and out of our gardens, nitrogen rarely stays in one place or one form. This has to do with soil chemistry.

Soils actually have an electric charge. In most climates, it's a negative one. Luckily, most soil nutrients have positive charges. In nature opposites attract, which means the positive nutrients stick to the negatively charged soil. Most soil nitrogen, on the other hand, is negative. Because of this, most nitrogen can't really stick to the soil. Instead, it drains out rapidly with soil water. When that water drains out of the soil with a big rain or irrigation, nitrogen is lost forever. The quicker your soil drains, the faster you lose nitrogen. If you add more nitrogen than the garden soils and plants can use, you are literally pouring it down a drain that leads to natural rivers, streams, and aquifers.

Nitrogen is also lost to the air in a process called volatilization. This happens when you spread a nitrogen source, such as manure or fertilizer, on the soil surface. Up to 50 percent of nitrogen in these materials can escape to the atmosphere if left on the soil surface for twenty-four hours. Because of this, it is best to mix nitrogen-rich material into the soil as soon as possible.

To reduce nitrogen losses, we add nitrogen when plants actually need it, during their fast early-growth period. Sandy soils, which don't hold on to soil water, are particularly vulnerable to losing nitrogen through leaching. Adding nitrogen in two or three doses throughout the season, rather than a single big dose, prevents loss and provides a sustained nutrient supply.

Because it leaches easily, don't fertilize with nitrogen or nitrogen-rich organic materials right before a big rain, and don't apply a heavy irrigation afterward. Nitrogen can also easily leak out of composts, manures, and other piles of organic materials, which makes them far less valuable as soil food. Be sure to cover compost and manure piles to keep this from happening.

Nitrogen is also unique in the way that it enters the soil. It is the only nutrient that plants literally pull out of thin air. Well, it's not the plants exactly, but nitrogen-fixing

Nitrogen-fixing bacteria grow in nodules on the roots of plants in the bean family (legumes). They harvest nitrogen from the air and pull it into the soil to feeds plants and microbes. Magnification 40X.

bacteria that live in the roots of certain plants. Most of these nitrogen-fixing plants are legumes (members of the bean family). When these plants shed roots and leaves, or when they are tilled into the soil, they actually enrich the soil in the nitrogen that they took from the air. For the soil-based gardener, including these plants—as part of a crop rotation, a cover crop, an intercrop, a perennial border, or a hedgerow—provides a steady and free source of nitrogen-rich material to feed the soil.

Whether it's added by fertilizer, mixed in through organic material, or captured out of thin air, the way we keep nitrogen in the soil is by making it part of the living soil. Soil organisms are made of nitrogen. By taking nitrogen into their living bodies they keep, recycle, and even capture nitrogen. In this way, you can think of your living soil as a nitrogen sponge, maximizing soil nutrients and their availability over the entire growing season.

At the end of the growing season or between crop rotations, a cover crop can also act like a nitrogen sponge. By taking up the leftover nitrogen in the soil solution, the cover crop holds onto nitrogen and keeps it from leaching away. When the cover crop is later incorporated, that nitrogen is again recycled and slowly released for the next season or rotation.

THE GOSPEL OF NITROGEN

Essential rules for keeping nitrogen in the soil:
1. Don't add more nitrogen than your plants need.
2. Add nitrogen in small doses and split applications.
3. Time nitrogen availability to match plant needs.
4. Don't pour nitrogen down the drain; avoid fertilizing right before a big rain or irrigation.
5. In areas of winter rain, fertilize in the spring.
6. Use cover crops to take up extra nitrogen at the end of the season or between crop rotations.
7. Keep composts and manures covered.
8. Don't make nitrogen volatile. Incorporate manure or other nitrogen-rich material into the soil within twelve hours of spreading.
9. Capture free nitrogen from the air using nitrogen-fixing plants as cover crops, crop rotations, or intercrops.

A MAGIC TRICK: PLANTS THAT PULL NITROGEN OUT OF THIN AIR

Including nitrogen-fixing plants in the whole-garden system builds sustainability by capturing nitrogen from the atmosphere. Add nitrogen to the soil by intercropping these plants, turning them in as green manures, adding them to composts, or collecting leaves and trimmings from perennial trees and shrubs. This is a partial list of nitrogen fixers that can be incorporated into a whole-garden system.

Garden Plants	Green Manures and Cover Crops	Perennials	Trees and Shrubs
Fenugreek	Alfalfa	Ceanothus	Acacia
Garden peas	Clover	Licorice	Alder
Peanuts	Cowpea	Lupine	Goumi berry
Pole beans	Fava beans	Milkvetch	Locust
Soybeans	Vetch	Wisteria	Silverberry

Chip trimmings from locust-planted hedges or windbreaks to make nitrogen-rich compost or mulch.

White clover adds nitrogen when interspersed in lawns, as living pathways, or as green manures.

When interplanting the classic three sisters—corn, beans, and squash—pole beans contribute nitrogen to the growing guild.

Lupine in the perennial garden sheds roots and leaves to add nitrogen to neighboring plants.

THE SOIL-NUTRIENT TEST

A soil test is the first step in finding out what important nutrients your soils are missing. From this starting point, you can best figure out how to correct these deficits. Once soil nutrition is balanced, the living soil has the potential to supply your garden's nutrient needs. In the beginning, however, your soil may need a jump-start.

A soil-nutrient test also lets us know if we have too much of a nutrient. Excess nutrients cause garden problems or, at the very least, lead to a waste of resources. I, personally, don't want to spend money or time adding uneccesary fertilizer. What's more, excessive fertilization often causes water pollution.

For a meaningful soil test, send samples to a professional soil-testing laboratory. Choose a soil lab that is certified by the North American Proficiency Testing Program. You can find a list of certified labs on the NAPT website (www.naptprogram.org). To decide on a lab, call around for prices and what tests the labs offer. Ask for a copy of a sample test report to see if it makes sense to you. As results differ slightly from lab to lab, it is important to keep using the same soil lab in subsequent years to compare reports over time.

Uncertified labs generally do not have the resources to provide a quality and verifiable soil test. Soil home-test kits are available but not very accurate. If you use a home kit for a quick and dirty check, be sure to follow the directions carefully, including using distilled water to mix in your soil sample.

When to Test

Testing soils before starting a new garden is critical, as soil nutrient deficiencies are easily corrected while preparing new garden beds. For new perennial landscapes, including lawns, fruit trees, or ornamentals, this is particularly true, as they require a big up-front investment. In all cases, providing the right soil nutrients from the get-go is essential for a successful garden and a satisfied gardener.

For established gardens, lawns, or trees, a soil test is a great check on maintaining and building soil fertility over time. Test vegetable and other annual gardens every two to three years and perennial gardens, trees, and lawns every three to five years. These periodic tests let you correct imbalances from over- or underfertilizing. Because we don't often know the compositions of organic residues or manures, a soil test can prevent unintended nutrient buildup or deficits in the soil.

Soil tests are best taken at the end of the summer or in the fall before the next year's growing season. This allows us ample time to add amendments and fertilizers while avoiding the spring rush on soil-testing labs. Getting early results from a fall soil test is particularly important when using lime and organic amendments. These materials can take months to break down and become available.

How to Test

A soil test is only as good as the samples that you take. The less than 1 pound of sample that you send to the lab represents literally tons of soil. Because of this, it is good to take some time to think about the sample. Taking a soil test has three elements:

Where to Test

Where to take a soil test depends on your objectives. If you have several different, sizeable elements to your garden, then ideally, you would test each of these separately. For instance, since they have different nutrient needs, I may take separate tests for my vegetable garden, my fruit trees, my landscape garden, and my lawn. This also takes advantage of the recommendations soil-testing labs will provide based on what you are growing.

Another reason to divide your garden into separate testing areas is if you notice obvious differences in soil characteristics or plant health. These differences across your garden could occur because of adjacent different soil types, areas with different histories, or areas with differences in slope or the water movement.

Garden size and practicality may limit you to a single test. If the soil is relatively uniform, a single test may also be used when establishing a new garden. After planting, you may choose to submit different soil tests for vegetables, perennials, and lawns.

How Many Samples

After deciding how many separate soil tests to take, look at each of these soil-testing areas as a unit. For each unit, you want to take five to ten subsamples that you will mix together to send off to the soil lab. Taking multiple subsamples gives a better average representation of the soil-testing area. It avoids the risk of sampling only the unique spots in the garden. For instance, the soil where your dog recently peed, though exceptionally fertile, would considerably skew your soil test results. When sampling, avoid garden edges as well as unique features, such as fence posts, garden borders, or paths.

To get the most representative sample, collect subsamples randomly. This is done by walking across in a zigzag pattern and sampling every few steps. Alternatively, divide the garden area into a grid and take subsamples at preselected points.

How to Collect Samples

To take a soil sample, you will need a bucket, a spade or trowel, a ruler, plastic bags, and a black marking pen. Use clean sampling tools and avoid galvanized metal, brass, or bronze metal coatings, as these can affect micronutrient results.

1. For each subsample, first scrape away the top layer of mulch, living plant material, or fresh organic matter. This should expose the bare mineral soil.

For preplanting tests when establishing new gardens, or for testing annual gardens, use the spade or trowel to dig about 6 to 8 inches below the soil surface. For established perennial gardens, use the trowel to dig about 2 inches below the soil surface. Use the ruler to check your depth.

2. With the spade or trowel, evenly scrape a slice off the side of the hole that you dug, trying to get an even amount of sample from the entire depth. Place the entire sample in the bucket.

3. Collect the remaining random subsamples in the same way, careful to make sure that you are sampling from the same depth each time. Place all subsamples from the same testing area into the bucket.

4. Mix all the soil subsamples in the bucket with the trowel.

5. Follow the instructions from your soil lab for packaging and labeling. Generally, you will place a large handful of soil from the bucket into a labeled plastic bag. Regardless of the labeling that the lab requests, be sure to keep a record of when, where, and to what depth each sample was collected.

Follow this procedure for any other soil-testing areas that you selected. Each testing area will result in one labeled sample bag to send to the soil laboratory.

What to Test

Most soil labs offer a standard menu for a basic soil-nutrient test. This generally includes soil texture, soil pH, organic matter content, and essential macronutrients. A standard soil test may also include micronutrient analysis, or you may need to specifically request it. If just starting out, it doesn't hurt to get the suite of micronutrients analyzed. In other cases, you'll want to test for micronutrients if you suspect a deficiency.

Sample/Field Number: MARK

SOIL TEST RESULTS

Estimated Soil Texture	Organic Matter %	Soluble Salts mmhos/cm	pH	Buffer Index	Nitrate NO3-N ppm	Olsen Phosphorus ppm P	Bray 1 Phosphorus ppm P	Potassium ppm K	Sulfur SO4-S ppm	Zinc ppm	Iron ppm	Manganese ppm	Copper ppm	Boron ppm	Calcium ppm	Magnesium ppm	Lead ppm
Medium	8.3		7.2				100+	188									

INTERPRETATION OF SOIL TEST RESULTS

Phosphorus (P) PPPPPPPPPPPPPPPPPPPPPPPPPPPPPPPPPPPPPPPPPF

5   10   15   20   25
Low   Medium   High   V. High

pH **
3.0   4.0   5.0   6.0   7.0   8.0   9.0
Acid   Optimum   Alkaline

Potassium (K) KKKKKKKKKKKKKKKKKKKKKKKKK

25   75   125   175   225
Low   Medium   High   V. High

Soluble Salts
0 1.0 2.0 3.0 4.0 5.0 6.0 7.0 8.0 9.0 10.0
Satisfactory   Possible Problem   Excessive Salts

RECOMMENDATIONS FOR: Vegetable garden

LIME RECOMMENDATION: 0 LBS/100 SQ.FT.
TOTAL AMOUNT OF EACH NUTRIENT TO APPLY PER YEAR:*

NITROGEN
0.15 LBS/100 SQ.FT.

PHOSPHATE
0 LBS/100 SQ.FT.

POTASH
0.1 LBS/100 SQ.FT.

THE APPROXIMATE RATIO OR PROPORTION OF THESE NUTRIENTS IS: 30-0-20

Use a fertilizer with the percentage of nutrients closest to the above ratio. Apply according to the instructions on the fertilizer bag or container, or determine the amount required from the instructions given on the back side of this report. Since meeting the exact amount required for each nutrient will not be possible in most cases, it is more important to apply the amount of nitrogen required and compromise some for phosphate and potash.

If a fertilizer contains phosphate and/or potash, it can be mixed in the spring or fall into the top 4-6 inches of topsoil. If a fertilizer containing only nitrogen is used, it should be applied in the spring, tilling or raking it into the surface. Nitrogen is easily leached through soil.

For sweetcorn, tomatoes, cabbage, and vine crops such as squash and cucumbers, an additional application of 1/6 lb. nitrogen per 100 sq. ft. may be desirable at midseason. This can be accomplished by applying 1/2 lb. (about one cup) of 34-0-0 fertilizer. Throughly water fertilizer into the soil.

A soil test report comes in a variety of formats. The most basic test reveals whether you have low, medium, or high levels of essential plant nutrients.

SOIL TEST RESULTS: WHAT DOES IT ALL MEAN?

The results of your soil-nutrient test will return from the lab as a soil-test report. Different labs have different formats for this report, but, in general, a report has three basic elements: the measured amount of each nutrient, the lab's interpretation of whether this amount is low or high, and the lab's fertilizer recommendations based on what you are growing. Now, it's your job to interpret what all these numbers and letters mean to you.

When a lab tests soil nutrients, its workers measure what is readily available to the plants—that is, the nutrients dissolved in soil water. As we build organic matter and create living soils, most of soil nutrients become tied up in the organic part of the soil, slowly released as this decomposes. Because of this, the nutrients measured by the test report represent only a fraction of the potential nutrients available in your soil. Lab fertilizer recommendations are generally much higher than necessary. Nonetheless, the test report is a good starting point, as long as we realize its limitations. A nutrient deficiency detected by the soil test report likely indicates there is deficiency in the living portion of the soil as well.

Soil Properties

A full soil test report will provide useful information about soil properties in addition to nutrients. Useful soil test reports will measure soil texture and provide the actual percentages of sand, silt, and clay. From this information, you can use the USDA textural triangle from Chapter 1 to determine your soil type.

Instead of percentages, some tests give soil texture ratings of very fine, fine, medium, coarse, and very coarse. This is basically a scale of how much clay (fine) versus sand (coarse) is in your soil. A medium texture rating indicates that you have a good mix of sand and clay, the flexible and forgiving loamy texture described in Chapter 1.

Organic matter is reported as a percentage. When measured by the same lab over time, this is a useful tool to evaluate how well we are building soil health. Though there is an upper limit of organic matter in a soil, it should increase during the stage when we are building soil health. Generally, a soil with greater than 5 percent organic matter is considered very high. A low rating means there is a long way to go (and a lot of potential) to build organic matter levels in your soil. This number is also used as a starting point in Chapter 5 to estimate the nutrient-supplying power of the organic portion of your soil.

Soluble salts are also reported on the soil test. Highly soluble salts lead to chronically bad soil structure and can be toxic to soil and plant life. These soils are difficult to revitalize, don't hold on to nutrients, and impede plant growth and seed germination. Arid climates are most at risk for salt accumulation, but salts can build up with excessive fertilizers and manures, even in humid regions. If you use manure as an amendment, this is something to watch on periodic test reports. If levels start to rise, it's time to switch to other amendments. Consult a specialist to reclaim soils with high salt levels.

Nutrients

Nutrients are reported in parts per million, or ppm. Because this number can vary from lab to lab, depending on soil testing methods used, I generally rely on the lab report's interpretation of whether this is a very low, low, medium, high, or very high nutrient level. A nutrient with low or very low levels is likely deficient in the soil. In these cases, I will consider fertilizing to correct the problem. For medium levels, I will consider what is available from organic matter. At high or very high levels, I avoid adding fertilizers or amendments that are concentrated in those particular nutrients.

Nitrate—Nitrogen

Nitrogen—the holy grail of plant productivity—comes in many forms. Much of it is tied up in the bodies of soil organisms and organic matter. A soil test, however, measures only the readily available nitrate in soil water. Since most of this soil water drains away, the soil test report for nitrogen is not very useful; it doesn't give a good account of the nitrogen truly available in a living soil. For the same reason, it doesn't give a good indication of how much nitrogen fertilizer to add. In fact, a soil report will often give fertilizer recommendations based on the N requirements of a plant without considering the soil's supply.

Counterintuitively, we want to see a low test result for nitrate-nitrogen, particularly at the end of the growing season. This means you've done a good job of capturing and keeping nitrogen in the living soil. When plants and soil microbes absorb all the added nitrogen, it is not lost from the soil. High levels of nitrate-nitrogen at the end of the growing season, however, means that more nitrogen was added than plants and microbes could use. If this happens, a postseason cover crop can catch excess nitrogen and keep it in the garden ecosystem.

Phosphorus

The P of NPK refers to phosphorus. Phosphorus is essential for all the energy-requiring processes in living things. For plants, these activities include photosynthesis, growing new root and stem tips, and setting flowers and fruit. Since phosphorus is so important for initiating growth, it is particularly important for young seedlings and transplants.

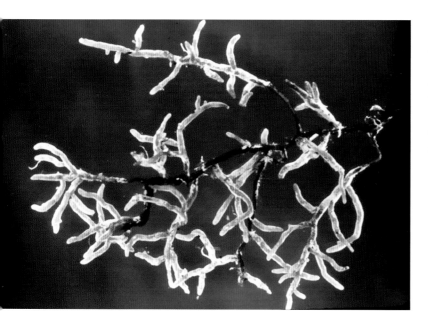

Mycorrhizal fungi grow on the tips of plant roots. They extend miles of threadlike fungal fingers to reach deep into the soil. These mycorrhizal networks are important in mining phosphorus from rocks and delivering it directly to plant roots.

In nature, phosphorus is weathered from rocks in the soil. Phosphorus also makes up a large percentage of organic matter and is often concentrated in manures. When organic matter is decomposed by the living soil, organic phosphorus is released for use by plants and microbes.

Unlike nitrogen, phosphorus is not easily lost from the garden. Phosphorus sticks strongly and readily to soil particles. It sticks so strongly that it is sometimes not easily released to plants. Often due to this strong attraction, in fact, phosphorus is the most commonly limited garden nutrient after nitrogen.

On the soil-test report, phosphorus results have two values: the first is used when the soil's pH is under 7, and the second is used when the pH is over 7, usually in dry climates. Your soil report should tell you which one to use. The phosphorus results are actually fairly good indicators of what we have and what we need to add.

Since phosphorus is limited by sticking to the soil, the strategy for making it more available to plants is to build high levels of soil phosphorus, like keeping the oil in your car topped off so that there is more than enough for the engine to use. This is done by adding plenty of organic matter and the occasional phosphorus fertilizer in the early stages of garden establishment. Because soil phosphorus is not easily lost, it is continually recycled between plant and soil. It is taken up by living plants and returns to the soil as organic matter when they die. In this way, once soil phosphorus supply is built, it naturally maintains relatively stable levels as long as organic matter is continually replenished.

Previously fertilized soils in old gardens or agricultural lands often have high levels of soil phosphorus. In this case, draw down soil phosphorus. At high or very high phosphorus, avoid phosphorus-rich material, including animal manures. Similar to nitrogen, too much soil phosphorus causes water pollution. Consequently, unnecessarily adding phosphorus is not only a waste but causes environmental problems as well.

Potassium

Potassium, the K in NPK, is the third most critically managed nutrient in soils. In prior days, it was called potash, a word you might come across from time to time in garden books or on fertilizer bags. Unlike most other nutrients, plants don't experience toxicity with too much potassium. In fact, if it is available, plants will continue to consume potassium once they are full. Because of this adding more potassium fertilizer than the plants actually need is a waste of money and resources.

Potassium is a salt. An electrolyte sports drink is filled with potassium and other salts to prevent dehydration. Similarly, in the soil, potassium is responsible for all the water relations in a plant. This leads to some very important functions, such as regulating

Potassium Deficiency

Potassium controls all the water relations in a plant. Leaves with crisp, burnt-looking edges are symptomatic of potassium deficiency.

gas exchange and nutrient uptake, as well as controlling the juiciness of a tomato, the crispness of an apple, or the sweetness of a carrot. Potassium also helps plants deal with stresses, such as drought, pests, and excess nitrogen.

Like phosphorus, potassium naturally exists in soil rocks. With its positive charge, it sticks around and builds up in the soil. Potassium is generally plentiful and plant-available. Since it is not easily lost, it is readily recycled and released by organic matter.

Soil tests give reliable results for available soil potassium. For soils rich in organic matter, consider potassium fertilization at low or very low test-report levels. As with phosphorus, draw down potassium levels when they are high by avoiding fertilization.

Calcium, Magnesium, and Sulfur

The remaining three macronutrients—calcium, magnesium, and sulfur—are usually sufficiently available in most soils. Shortages can occur, however, depending on the region or climate. In addition to doing a soil test, check with local garden centers or agricultural extension experts about whether these nutrients are commonly limited in your area.

Of these three nutrients, sulfur is the most likely to be deficient, particularly in sandy soils. Like nitrogen, sulfur is negatively charged in soils and easily drains with soil water. Sulfur is readily available from decomposing organic matter. For this reason, living soils that recycle and retain organic matter rarely show sulfur deficiencies.

Until recently, sulfur was abundantly supplied by industrial air pollution, known as acid rain. In fact, it has only been in the last few decades with successful environmental cleanup that sulfur deficiencies have become common across the United States. Because of its increasing prevalence as a limiting garden nutrient, fertilizer bags have started listing four dashed numbers on their label instead of the usual three. In this case, the fourth number represents the amount of sulfur, in addition to NPK, contained in the fertilizer bag. Sulfur, reported as sulfate on the soil test, should be interpreted similarly to

Blossom-End Rot

Blossom-end rot, as shown on this tomato, is a common symptom of calcium deficiency. It can also be caused by improper watering in dry areas.

nitrate-nitrogen. Combining soil test information and an estimate of the organic matter supply will determine if you need to add this nutrient.

Calcium and magnesium are readily supplied by soil rocks. These positively charged nutrients stick around in the soil once they are added. For chemical reasons, buildup of these nutrients improves overall soil structure and the nutrient-supplying power of the soil.

The test report is pretty reliable when it comes to calcium and magnesium. If reported as low, you'll want to find a way to add it. For acidic soils (pH of less than 7) you can add calcium by using lime and calcium and magnesium by using dolomite lime. If soils are neutral to alkaline (pH of 7 or more) and don't need liming, then add calcium using gypsum, an amendment that also supplies sulfur.

Micronutrients

Micronutrients are required by plants in really small quantities, yet they are still essential for plant growth. A deficiency in the minute amount of micronutrients required by your garden can have drastic consequences for the health of your plants. For instance, without a tiny amount of manganese, plants can't build or grow new leaves, stems, or roots. When this happens, nitrogen and phosphorus build up in plant tissues, leading to root and leaf diseases.

Clods of heavy clay soil can make gardening difficult. Gypsum softens these soils and improves structure. As an added bonus, this soil conditioner supplies calcium and sulfur.

Interveinal Chlorosis

The yellowing of leaf veins (interveinal chlorosis) is a common sign of iron deficiency. Iron deficiency commonly occurs in high-pH (alkaline) or waterlogged soils.

At the same time, too much of any of these micronutrients are toxic and even fatal for plants. Because they are needed in such small quantities, the difference between toxic and deficient levels is a fine line. Care should be taken not to over-apply any of the essential micronutrients. If you do choose to fertilize with micronutrients, be sure to follow recommendations exactly to avoid under- or over-application.

Luckily, most micronutrients are well supplied by soil rocks or decomposing organic matter. Deficiencies may occur when micronutrients are naturally lacking in soils in your region or when soils have been excessively cropped and degraded. Micronutrient deficiencies also occur at low and high pH (discussed below).

There are a number of natural and organic amendments, such as seaweed-based materials, that are rich in micronutrients. These can be used to supply low doses of these necessary minerals. Some gardeners do this as a preventative measure to make sure all plant nutrient needs are met before the season starts, while others use these organic materials during the season only if they suspect a micronutrient deficiency.

Keeping It in Balance: pH and Lime

Soil pH is a measure of the relative acidity of soil on a scale of 0 to 14. A pH of 0 to 7 is acidic, from slightly so (pH 6.5) to melt-a-metal-container acidic (pH 0), while a pH of 7 to 14 is alkaline, from just a little (pH 7.5) to cause-serious-skin-burns alkaline (pH 14). Lemon juice is an example of fairly acidic substance, while bleach is an example of an alkaline one.

Some plants, such as blueberries and azaleas, are particularly acid loving. They thrive in low-pH soils (around 5.5) and are intolerant of alkaline conditions. These plants do well in sandy soils.

Plants and soil organisms have a preferred optimum pH that is generally at or slightly below neutral. Many plants are tolerant of acidic or alkaline conditions outside of their optimum range and can do well from pH 5.5 to 7.8. Others are highly sensitive or may need particularly acidic or alkaline soils. For instance, grasses (including lawn grasses) are tolerant of fairly acidic pH, while clovers are tolerant of fairly high ones. This information helps you choose plants adapted to particular soils. Check your gardening guide for the optimum pH and pH tolerance for plants of interest.

Soil texture plays a role in soil pH. Sandy soils tend toward acid, while heavy clay soils tend toward alkaline. Though sandy soils respond quickly, repeated liming is necessary to raise and maintain pH. It is difficult to manage the pH of heavy clay soils. In the case of clays, choosing plants tolerant of existing pH levels can bring greater success. Since both organic matter and fertilizers are acidic, most soils tend to acidify over time.

Proper soil pH is extremely important for plant nutrition. At either end of high or low pH, nutrients, particularly micronutrients, become unavailable to plants. At these extremes, toxins also become more abundant. For this reason, keeping your soil pH in balance is essential to providing good fertility for your living soil.

A laboratory soil test is the most accurate way to test pH. Usually, two results come back, labeled "soil pH" and "buffer index." Generally speaking, the "buffer index" is a

QUICK AND DIRTY CALCULATIONS FOR YEARLY LIME REQUIREMENTS

If your soil pH is ...	And your soil texture is ...		
	Sand	Loam	Clay
	... you need about this many pounds of *lime* per 100 square feet of garden*		
4.5	12½	25	35
5.0	10½	21	29
5.5	4	8½	11½
6.5	1½	3	4½
If your soil pH is ...	And your soil texture is ...		
	Sand	Loam	Clay
	... you need about this many pounds of *sulfur* per 100 square feet of garden*		
7.0	½	1	1½
7.5	1	2	2½
8.0	1½	3½	4½
8.5	3	5	6

Adapted from *Fertilizing Garden Soils* (Cornell Cooperative Extension, Chemung County).
*These requirements are based on the top 6 inches of garden soils. For perennials, divide this number by 3 to get the amount needed for the top 2 inches of soil.

more true reading of soil pH because it accounts for soil texture effects. The advantage of a soil test is that the lab will interpret these numbers and the lime required to adjust the soil pH. Just remember that the optimum and tolerance pH for most plants occurs over a broad range. Test reports, on the other hand, often provide lime requirements based on adjusting pH to an optimum, neutral pH value of 7. For this reason, depending on what you are growing, you may choose to use slightly less lime than that recommended by the soil test report to reduce lime costs, yet stay within your plants' range of tolerated pH values.

Do-it-yourself methods for soil pH testing include hand-held pH meters, color test strips, and pH probes, which give results with varying accuracy. I generally get a lab test for soil pH, but use a pH meter to monitor soils of particularly sensitive crops more frequently. It is important to carefully follow the directions for these pH monitoring products, as operator error can result in bad readings. Since these tools measure only soil

STEP-BY-STEP: THE pH BALANCING ACT

1. **Measure soil pH.** Do this when performing soil-nutrient tests at the end of the growing season. If possible, independently measure pH for vegetables, perennials, lawns, and acid-loving plants, as target pH changes depending on what you are growing.
2. **Choose a lime material.** Lime materials come in varying grades from fine powders to coarse granules. The finer the material, the more quickly it will react to raise the pH of your soil. The coarser the material, the longer lasting the effect, providing neutralizing power as it dissolves over several growing seasons. Choose coarse materials for sandy soils, which need a longer-lasting effect. Also choose a lime material to address nutrient deficiencies, if any. Lime (calcium carbonate) adds calcium to the soil, while dolomitic lime adds calcium and magnesium to the soil.
3. **Determine the amount of lime to add using tables or soil test result recommendations.** This amount depends on soil pH, target pH, and depth. Place lime in soils to a depth of 2 inches for established perennials, trees, and lawns and to a depth of 6 inches for vegetable gardens and new perennial gardens or lawns.
4. **Add lime at least one month before planting, as it needs time to react with the soil.** Adding lime in the fall before the next growing season gives the best results. If adding other fertilizers or amendments, lime can be spread with them and incorporated to a 6-inch depth for vegetable gardens or new perennial, tree, or lawn plantings. For established perennials, spread lime on the soil surface and gently scratch into the top inch of soil using a rake. For established lawns, simply sprinkle lime on the soil surface.

pH (not buffer pH), estimate lime or sulfur requirements based on soil texture by using the tables on page 71.

Add lime in the fall before each growing season for annual gardens and during the soil preparation for lawns, trees, and perennial gardens. Acid soils are adjusted upward toward neutral pH using limestone (calcium carbonate). With magnesium-deficient soils, dolomitic limestone (calcium magnesium carbonate) also adds magnesium while raising pH. For alkaline soils (high pH), lower pH with elemental sulfur (certified organic is available) or a chemical fertilizer that contains sulfate, such as ammonium sulfate or iron sulfate.

For acid-loving plants, such as blueberries and azaleas, peat moss was traditionally used to acidify soils. Peat moss, however, is a nonrenewable resource that is mined from peat bogs at an unsustainable rate. As peat bogs are one of the biggest global reservoirs of carbon, mining this material has a huge climate change footprint. Compost, combined with acidic materials such as pine needles or coffee grounds, is a better, more environmentally friendly alternative to naturally acidify soil.

ENVIRONMENTAL SOIL TESTING

Occasionally, soils need a second type of laboratory test. Environmental soil tests check for contamination by toxic chemicals. In most cases, this type of test is unnecessary. For urban soils with unknown histories or soils possibly exposed to a contamination source, it is better to be safe than sorry. Eating food grown in or having direct skin contact with contaminated soils can pose potential health hazards, particularly for children.

Assess the risk of contaminated soils by first looking at the history of your garden site. Start by talking to neighbors or checking into city records. Investigate soils for contamination if your garden hosted industrial, mining, waste disposal, or manufacturing activities. Old railroad lots, gas stations, parking lots, manure storage sites, or fertilizer and pesticide storage sites are also at risk for contamination. It's a good idea to check for lead in soils surrounding older houses, as lead paint can shed and contaminate soils.

Once you determine your potential contamination source, you can identify what types of tests to request. Unfortunately, environmental soil tests are contaminant-specific. They are also quite expensive. If you are generally concerned about lead from paint or gasoline, a number of labs test this for a relatively low price. Sample your soils in the same way that you would for a soil-nutrient test. Send in separate tests for separate areas of concern. For other contaminants, consult a qualified professional to determine what to test, where to test, and how to interpret the risk.

CHAPTER 4

FOOD, WATER, SHELTER, AIR

ORGANIC AMENDMENTS TO GROW LIVING SOILS

Once we know the requirements of the living soil—food, water, shelter, and air—our job is to build gardens that naturally meet these needs. The best way to do this is with organic material that feeds and houses soil organisms.

When we add material to a soil to improve its condition, we call it an amendment. Amendments provide nutrients, improve soil tilth, increase biological activity, hold on to water, or balance pH. Bulky organic amendments, which build organic matter in the soil, accomplish multiple purposes at once. Because of this, using organic amendments is the foundation of any soil-growing program.

There are many ways to amend the soil with organic matter. These include trucking in manure from the neighbor's farm or adding piles of well-rotted compost. It also includes putting organic material grown from the soil back in the soil by turning in residues and planting cover crops. By following the guiding principles from Chapter 2—use what you have and give more than you take—it's easy to use organic amendments to build and replenish healthy soils.

Sometimes, especially in the early years of soil building, organic materials alone will not meet all of a soil's nutrient needs. In this case, it is essential to use concentrated fertilizers to supplement organic amendments. Chapter 5 specifically addresses how and when to use concentrated fertilizers as part of a whole-soil fertility plan.

WHAT IS ORGANIC?

When we talk about gardening, the word *organic* has two meanings. The first is a legal definition that describes certified-organic agriculture. This meaning refers to farms and gardens that use only fertilizers, amendments, and pesticides derived from natural sources and that are approved by an organic certification organization.

The second definition refers to a core principle of growing great soils: soil organic matter. In this case, organic matter refers to the impotant fraction of soil that is living or was once living and since transformed into the rich, black spongy gold that gardeners covet. We build this gardening treasure by adding organic materials to the soil, that is, bulky residues from once living plants and animals.

Building organic matter in soils does not depend on using all certified-organic materials. You can easily use all organic fertilizers and amendments and continue to deplete your soils. By the same token, a gardener who uses some synthetic material, like a chemical fertilizer, can also build great soils at the same time if they are focused on increasing soil organic matter. The ultimate goal in building great soil is to rely on the soil system itself, reducing outside inputs, whether organic or not. Because building soils relies on natural processes, we generally use natural materials, often certified organic, to accomplish soil-building goals. In this way, we work with, rather than against, the living soil community.

Two different definitions of organic: Certified-organic products are grown with only certified-organic materials. Organic amendments are bulky materials, such as animal manures, that come from a living source.

One sure way to work against the living soil is through the use of pesticides. Both organic (naturally derived) and nonorganic pesticides have extreme and potentially long-lasting effects on soil organisms by decimating whole populations of critters in the soil. Occasionally, a garden pest may warrant such extreme measures, but if at all possible, it is best to rely on natural soil-building techniques. Over time, the healthy and diverse soil biological community will work to keep in check undesirable weed, disease, and insect pest problems.

First and foremost, however, we'll dive into the what, when, and how of using the organic amendments that form the basis of any soil-building plan. The following sections cover how to choose, how to make, and how to use this vast array of materials—cover crops, plant residues, composts, manures, and yard debris. These tools turn garden waste into a soil feast, building healthy soils that become more and more productive over time.

THE RIGHT CHOICE

Choosing the right amendment is a matter of knowing our gardens and ourselves. Soil-quality assessments, soil-nutrient test results, and our own goals guide our choice of amendments. We'll choose different options depending on whether we want to add nutrients, build soil organic matter levels in the long term, or address a particular problem, such as compaction. The most important question that I ask myself, however, is "What can I get most easily, most of the time?" Because building organic matter levels requires consistently adding organic materials, access and availability are my primary concerns.

If your neighbors board horses, then free horse manure may be your go-to material. If you can't transport truckloads of compost, then growing a green manure in place is an ultra-local option. Urban and suburban areas offer a consistent waste stream of food refuse from restaurants or grass clippings from neighbors.

The types and choices of amendments for today's gardener are only limited by the imagination. There are lawn clippings and fall leaves, kitchen scraps, mushroom compost,

WHY ORGANIC MATTER?

The effects of organic amendments are lasting and dramatic, making your job as a gardener easier. Regardless of what they look like when they go in, manures, yard trimmings, tree leaves, or kitchen waste are equally transformed into rich, black spongy soil organic matter, the backbone of living soil. Organic amendments are your frontline strategy to improve the following:

- Structure, softness, and tilth (gives homes to soil organisms, gives space for plant roots to grow, saves your back from digging)
- Compaction (fluffs up a hard soil)
- Moisture and water retention (a sponge that holds onto water)
- Aeration and drainage (water drains from the soil to let life in the soil breathe)
- Biological activity (good habitat and good food = a healthy and happy population)
- Nutrient availability, recycling, and retention (food warehouses for the living soil)

Creativity rules for using and choosing organic amendments. Use what's at hand, from tree trimmings to manures to kitchen scraps, to add organic matter and grow garden soils.

coffee grounds, cardboard, chicken droppings, straw and hay, wood ash, chipped-up blackberry canes, and even spent grain from your local brewery. If it was once living, and it doesn't contain any toxins, then you can add it to the soil.

Quality Matters: The Brown-to-Green Ratio Explained

Not all organic amendments are created equal. Sawdust and chicken manure both add organic matter, but there is a big difference in how they look, feel, and smell. This translates into differences in how they taste to soil critters.

Amendment quality describes these differences. In other words, how scrumptious is the organic feast from the living soil's perspective? Quite simply, quality refers to the amount of nitrogen in the organic material. The carbon-to-nitrogen ratio (C:N ratio), otherwise known as the "brown-to-green" ratio, measures this quality. Materials with a high C:N ratio (low nitrogen) have low quality, while materials with a low C:N ratio (high nitrogen) have high quality. The "brown-to-green" alias comes from the high-nitrogen green materials (such as fresh grass clippings) and high-carbon brown materials (such as wheat straw) that are used to balance a compost pile. This can be confusing since not all "green" materials are actually green. Fresh manures, kitchen scraps, nitrogen fertilizers, and coffee grounds are also examples of high-nitrogen "green" materials.

The C:N ratio determines how fast organic amendments are gobbled up by soil organisms. High-quality amendments are quickly eaten (decomposed) by microbes, rapidly releasing nutrients and rapidly disappearing. Low- to intermediate-quality amendments are less tasty to critters. They are eaten more slowly, which means they take a long time to release nutrients. In fact, microorganisms may actually use, instead of release, soil nitrogen to eat very low-quality organic food. Although less nutrient-rich, low- to intermediate-quality materials build up the organic part of the soil because they stick around longer.

When we feed the soil, we want to add materials with a range of qualities. High-quality materials quickly supply plant nutrients. Intermediate- to low-quality material builds organic matter. A very low-quality material can come in handy at the end of the growing season by locking up leftover nitrogen, which keeps it from draining away.

More Quality Considerations

Organic amendments are increasingly available through commercial or municipal composting programs. Manures and other materials can be bought in bulk. When buying these commercially available amendments, assess quality to make sure they meet your goals and don't contain any detrimental ingredients, such as unwanted pests, weed seeds, or excessive salts.

First, find out as much information as you can from the seller. If purchasing compost, ask what raw materials were used and how long they were composted. What temperatures

Low-quality materials include straw, bark, and wood chips. They have a high C:N ratio, break down very slowly, and can actually take nitrogen from the soil.

Intermediate-quality materials include compost and rotten leaves. They decompose slowly, release some nutrients, and build organic matter.

High-quality materials include nitrogen-rich fresh manures, young legumes, and fresh grass clippings. With a low C:N ratio, they decompose quickly to release loads of nutrients.

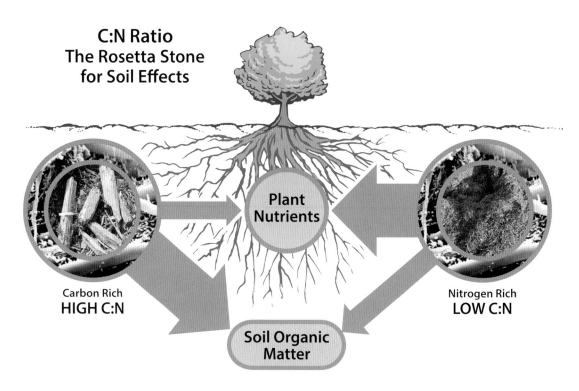

C:N Ratio
The Rosetta Stone
for Soil Effects

Plant Nutrients

Carbon Rich
HIGH C:N

Nitrogen Rich
LOW C:N

Soil Organic Matter

The C:N ratio is the Rosetta Stone for understanding amendment effects. High C:N (low-quality) materials are slowly decomposed by soil microbes, thus contributing more to building long-lasting organic matter in the soil. Low C:N (high-quality) materials are quickly consumed by microbes. These rich materials rapidly disappear and excess nutrients are made immediately available to plants.

did the piles reach and how was finished compost stored? This will give you an idea of the C:N ratio of the compost and whether it contains toxins or unwanted substances. Find out whether the compost was sterilized. Though a soil food source for your soil, sterilized compost is less beneficial. Although you may not get answers to all your questions, the more information available, the better the chance that you are getting a quality product.

When purchasing a commercial product, also use your own senses. I want to create a living soil that is messy, complex, and fertile; I want to use a messy, complex, and fertile amendment as well. When I buy compost, I touch it and smell it. If it doesn't look alive, it probably isn't. While my head is in the compost, I'll also check out color, odor, texture, and consistency. A good compost has a rich, black color; an earthy (not ammonia) smell; a relatively consistent, fine particle size; and more or less uniform contents from batch to batch.

For both composts and manures, I also want to make sure I'm not importing pests or weed seeds. Symphylans, a nasty soil pest, live in organic amendments in the western

CARBON-TO-NITROGEN RATIOS (C:N)

The carbon-to-nitrogen (C:N) ratio determines the quality and effect an organic amendment. Generally, materials with a C:N ratio of less than 20 release nitrogen to the soil. Materials with a ratio greater than 25 will initially take some nitrogen from the soil.

	C:N Ratio*	Quality
Bloodmeal	4	High
Buckwheat cover crop	34	Intermediate
Cardboard	500	Low
Coffee grounds	20	High
Corn stalks	60	Intermediate
Early alfalfa hay	13	High
Fish emulsion	8	High
Grass clippings	10–25	High
Homemade compost	16–30	Intermediate
Kitchen scraps	10–20	High
Leaves	40–80	Intermediate–low
Legume cover crop	10–20	High
Manure, fresh without bedding (varies widely)	5–30	High–intermediate
Mature alfalfa hay	25	Intermediate
Paper	100–200	Low
Sawdust	600	Extremely low
Shrub trimmings	53	Intermediate
Soybean meal	4–6	High
Tree trimmings	16	High
Wheat straw	100–150	Low
Worm castings	12–15	High

*Numbers are average estimates compiled from several different sources. Actual C:N ratios can vary widely.

United States. To check for these unwanted critters, I spread the compost on a black sheet and make sure no white bugs come crawling. I also ask neighbors and friends about new and unverified compost and manure sources. If not heated enough, composts and manures can contain weed seeds. If you're concerned, check for weeds by taking a small sample of the material, wetting it slightly, wrapping it in a moist paper towel, and putting it in an open Ziploc bag. Keep the compost slightly moist and in a warm place for several days to see if weeds germinate.

Smell, touch, and imagine how your compost tastes to soil organisms. The best compost is alive and full of rich, dark organic matter.

Amendment Quality Test

A laboratory test is another way to assess amendment quality. In addition to soil, most soil-testing labs also test amendment quality. Because manures and other amendments vary from batch to batch, this is actually the only way to accurately assess amendment nutrient content. For farmers and others whose income depends on this information, these tests are more than worth the cost. Commercial composters should also have these lab results available upon request.

An amendment test provides the following information to help you assess quality:

- **Nutrient content.** To provide plant-available nutrients, amendments should have a C:N ratio of less than 20:1.
- **Ash content.** An ash content greater than 50 percent or a bulk density greater than 1 means that compost composition is questionable and likely containing more mineral soil than organic material.
- **Alkalinity or acidity.** To avoid imbalances in soil pH, use amendments with a plant-compatible pH of 6.0 to 7.8. Avoid manures with excessively high pH.
- **Salts and sodium content.** Amendments high in salts can cause soil-quality problems. Use amendments with soluble salt less than 10 mmhos/cm and a SAR ratio of less than 13 percent. For soils with existing salt problems, avoid salt-containing materials altogether.

GROW IT, BUILD IT, OR BRING IT: A GUIDE TO ORGANIC AMENDMENTS

Green or brown, living or dead, the imaginative gardener has an endless array of ways to amend the soil with organic material. This section covers how to grow, make, and use a variety of organic amendments to build a living soil. In addition to the amendments described below, any practices that return organic materials to the soil, such as intercropping, crop rotations, or no-till gardening, are also ways of amending the soil. The guiding principle is to give more organic matter than you take from your gardens. Over time this will lead to more and more nutrients retained and recycled by the living soil.

Grow Your Own: Cover Crops and Green Manures

The best way to amend the soil is to grow the amendment right where you plan on adding it. This is exactly what we do when growing cover crops and green manures. Not only does this technique add nutrients and build organic matter in the soil, it relies on

Grow your own amendments using cover crops and green manures to add organic matter and nutrients to your soil. A quick crop of spring mustard takes advantage of a gardening gap to enrich the soil.

Phacelia green manure provides beauty as well as bounty. In addition to enriching the soil, phacelia's attractive summer blooms provide habitat and food for pollinators and beneficial garden insects.

the living system to do the work for us, a key principle for soil-based gardening. Growing green manures reduces costs and labor involved in making compost, hauling manure, or collecting organic materials. It also eliminates the risk of introducing unwanted pests, weeds, or diseases from these imported sources.

Using cover crops and green manures provides multiple benefits for the living soil. The terms *cover crops* and *green manures* are generally used interchangeably, but cover crops specifically refer to the benefits associated with covering bare ground, while green manures refer to the benefits of adding nutrients and organic matter when tilled into the soil.

As a soil cover, cover crops protect bare soil from erosion and suppress weeds. Additionally, they provide diverse habitat for soil organisms, beneficial insects (critters that reduce insect pests), and aboveground pollinators. In this way, they greatly increase the diversity of the above- and belowground garden.

When incorporated as a green manure, cover crops add both nutrients and organic matter to the soil. As with other amendments, high-quality (high-nitrogen) cover crops decompose quickly to fertilize the garden. These are generally legumes (bean-family crops), which fix nitrogen from the atmosphere. Growing and tilling under a legume cover crop adds enough nitrogen to reduce or eliminate your need for other nitrogen fertilizers.

Non-legume cover crops are lower in quality (lower nitrogen) and take longer to decompose. Though they release fewer nutrients, these crops are indispensable for building organic matter in the long term. Part of their nutrient-supplying power comes from the fact that they scavenge and recycle excess soil nutrients left at the end of a growing season.

The other huge benefit of cover crops comes from their roots. Remember the guiding principle, as above, so below? Cover crops and green manures add a massive amount of organic material belowground, through their living roots. These messy roots are like candy to soil organisms. They constantly leak sugars, making them a hot spot for living soil activity. Finally, cover crops with large roots or root systems can work wonders on breaking up a compacted or clay soil. In this way, they till the soil for you as you watch them grow.

Forage radish can replace your rototiller in breaking up a heavy soil. Let late summer roots penetrate compacted clays and decompose over the winter to loosen and soften garden soils.

Although cover crops do a lot of work on their own, they still need some garden care. This includes preparing seedbeds, watering, and weeding as they become established. Cover-crop seeds can also be expensive, depending on your source. However, when considering the enormous benefits cover crops have for building soils, and the labor, transportation, and fertilizer savings they offer, you'll find that they are well worth the price of seeds and a little garden care.

Choosing the Cover

Forbs, grasses, and legumes—oh my! These are the broad categories of cover crops available to use in building your soil. Grasses are annuals or perennials that build organic matter. Legumes are bean-family members that fix nitrogen. Forbs are flowering plants that provide a variety of benefits, depending on the species. Using a cover crop takes some planning and experimentation. The following lists the questions to ask when deciding which cover crop is right for you. Consult the cover-crop guides in the Additional Resources section for more specific information about each species.

1. **N-fixer or not?** Is your goal to supply nutrients or build organic matter? If replacing fertilizers by supplying nutrients is most important, then choose a legume (for example, clover, beans, vetch, or peas). If building organic matter is at the top of your list, grow a grass or a forb (for example, oats, rye, buckwheat, phacelia, or a brassica).

2. **Annual or perennial?** You can use perennial cover crops in an annual rotation or a year or two in advance of planting. The drawback is that perennials require extra effort to kill. If you are building soil the no-till way, then choose an easy-to-uproot annual.

3. **Climate.** Like any other plant, cover crops have optimal gardening zones. Some warm-season summer cover crops won't grow in the north, and some frost-sensitive species won't die off with winter in the south. Depending on your climate, you'll choose and use cover crop species in different ways.

4. **Planting window.** Are you looking for something to cover bare ground in the summer or between crop rotations? Then choose a warm-season crop that you can plant from early spring through summer (buckwheat, phacelia, soybeans, cowpea). Do you want a crop to cover the ground over the winter, supplying organic matter and nutrients for spring planting? Then choose a cool-season cover crop that you plant in late summer to early fall (rye, clover, forage radish, oats). Remember that planting windows for particular species change depending on your climate.

5. **Other benefits.** Are there other benefits you want to get from your cover crop or green manure? To improve compacted soils, forage radish will work wonders. To attract pollinators, plant a flowering forb.

You can also experiment with planting a combination of cover crops to get a variety of benefits. It's good to choose a combination that has complementary growth patterns. A late summer, early fall mixture of oats with field peas or rye with vetch combines the N-fixing power of a legume with the root-building effects of a grass. In these combos, the grass also forms a scaffold for the legume. A flower-power summer combination of buckwheat and phacelia provides the weed-suppression properties of quick- and dense-growing buckwheat to help the phacelia get established.

Cover crops work great in combination. Ryegrass forms a scaffold for nitrogen-fixing vetch to climb. Together, these garden helpers add nitrogen and building soil organic matter at the same time.

GREEN MANURES IN THE GARDEN		
Forbs, grasses, and legumes provide a wealth of green manure and cover crop choices. The following is a limited selection of some of my favorites.		
Forbs	**Grasses**	**Legumes**
Buckwheat Benefits: Organic matter. Suppresses weeds. Attracts pollinators. Quick growing and easy to kill.	Oats Benefits: Organic matter. Soil cover. Root biomass. Easy to kill.	Crimson clover Benefits: Nitrogen. Attracts pollinators. Quick growing and easy to kill.
Phacelia Benefits: Organic matter. Suppresses weeds. Attracts pollinators. Beautiful garden addition.	Annual Rye Benefits: Organic matter. Soil cover. Root biomass. Great for erosion control. Caution: difficult to kill.	Red clover Benefits: Nitrogen. Attracts pollinators. Slow-growing and shade-tolerant. Organic matter.
Forage radish Benefits: Organic matter. Soil tilth. Soil compaction. Suppresses weeds over winter. Powerful soil conditioner.		Fava beans Benefits: Nitrogen. Suppresses eeds. Organic matter. Soil compaction.

Planting the Crop

Next, consider where and when to plant the cover. You may think you might not have space for a cover crop. Even the smallest garden can find creative niches between rows, between spring and fall plantings, and at the end of a growing season. You'll be amazed at how quickly adding a cover crop here or there can build organic matter.

One of the simplest ways to begin using cover crops is between spring- and fall-planted vegetables. After harvesting early vegetables, plant a quick-growing warm-season crop, such as buckwheat or phacelia. A couple of weeks before you are ready to plant fall vegetables, hand pull or mow the cover crop, leaving it on the soil surface. You can transplant directly into this mulch. Alternatively, lightly chop and turn the residue into the soil. Plant into the new seedbed after it mellows for a week or two. Even if you are not ready for fall planting, be sure to pull the cover crop when in flower and well before it goes to seed.

During the growing season, use summer cover crops, particularly buckwheat, as a living mulch that competes with weeds for light and moisture. To do this, interseed the cover crop a third of the way through the vegetable's growing season. Avoid competition with your vegetables by choosing cover-crop companions that won't outgrow your crops.

Plant cool-season cover crops at the end of the growing season. These species put on growth in the fall, adding organic matter and nutrients when incorporated in the spring. Even if they winterkill, fall-planted cover crops suppress weeds during the garden fallow period. Fall-planted cover crops must be seeded at least four weeks before the first

Use the gardening gap between spring and fall vegetables to plant a succulent summer green manure such as buckwheat. Even a few weeks is enough time to get some good effects from this fast-growing forb.

hard frost to give them time to establish. Either prepare a new seedbed for the cover crop by raking cover-crop seed into the soil surface or oversow directly into the residue of late-summer crops. To do this, broadcast the cover-crop seed over late crops at a higher seeding rate than normal.

To get the most from your N-fixing cover crops, you'll want to inoculate the seeds before planting. Nitrogen fixation requires the presence of rhizobium bacteria. These bacteria occur naturally in soil but may not be at sufficient levels to jump-start your legumes. Inoculating your seeds ensures that these hardworking bacteria will be plentiful and powerful from the get-go.

Inoculants are powders that you can purchase from the seed company. Each inoculant is specific to a particular legume species, so make sure to use the product that is matched for your seeds. Soak the seeds in the inoculant overnight before planting.

Killing the Crop

Successful cover cropping depends on successful killing. To get the most nutrient and organic matter benefits, timing is critical. For maximum nitrogen, kill crops just as they begin to bud. In a crunch, you'll also get benefits, though reduced organic matter, from killing off a cover crop before flowering. Avoid letting a cover crop go to seed, as nutrient quality suffers and they can actually become garden weeds.

When to kill also depends on the timing of your next crop. You'll want to let the cover crop residue mellow in the soil for a couple of weeks before planting. This gives the green manure time to decompose and begin releasing nutrients. Nitrogen-rich legumes and succulent buckwheat are faster to decompose and can be replanted with the next crop sooner after cover-crop incorporation.

Plant favas in the late summer or early fall to give your garden a nitrogen boost. To maximize nitrogen supplied, kill them just as they begin to flower.

Depending on your gardening style and the cover crop species you choose, there are several options for killing and incorporating the green manure:

1. **Till it.** This is the traditional method of cover-crop incorporation. Tillage works best if you have a tractor or rototiller that can make several passes. You can also use a spade and fork to dig up, chop, and dig in your cover crop. For the small-scale gardener, and with a small-size rototiller, this can be fairly labor intensive.
2. **Winterkill.** This is the most effective way to kill your fall cover crop. If you are lucky enough to live in a winterkill zone (generally zone 6 and colder), frost-sensitive cover crops will die off and naturally decompose over the winter. Without lifting a finger, you'll have a richly amended garden bed that is ready to plant with the spring.
3. **Mow, pull, and mulch.** Mowing won't successfully kill all cover crops, but it's a good way to knock down certain summer-growing and shallow-rooted forbs, peas, and beans. Mown cover crops remain in the soil to build organic matter as they decompose. For crops that readily spring back, such as some grasses and clovers, pull them up by the roots. Once mown or pulled, leave the cover crop as

Pull up the quick-growing buckwheat cover crop by the roots and use it to mulch a new summer planting. It will protect new seedlings, while building organic matter as it decomposes.

mulch on the soil surface and plant directly into it. You can also pull the residue off the bed for planting and replace it as mulch when the new seedlings become established. The mulch will protect the new crop, while adding organic matter to soil.

4. **Sheet mulching.** Sheet mulching, explained in the next section, is a way to amend the soil. It's also a great way to kill cover crops that won't winterkill while avoiding the extra work of tilling them in by hand. Mow or chop the cover to the ground, layer with wetted cardboard, and add sheet mulch layers. Kill your cover crop with a sheet mulch two to three months before you want to plant into it.

Raw Power: Composting in Place

I'm not lazy by nature, but I do like to avoid unnecessary work. Because of this, one of my favorite ways to amend the soil is to simply add raw, uncomposted organic materials directly to the soil. These raw materials can include anything that would normally go into a compost pile: manure, grass clippings, kitchen scraps, garden residues, and so on.

Using raw organic materials, the living soil itself becomes the compost bin. By composting in place, the work of building, maintaining, and transporting a pile is eliminated. Nutrient losses are also minimized. Depending on how you apply the materials, this technique can also cover the soil as mulch that suppresses weeds and retains water. Above all, composting in place minimizes disturbance to maximize soil ecosystem function.

There are several ways to add organic material directly to the soil: layering them on the surface, tilling them into the soil, digging them into a trench or hole, or using a combination of these. As with any type of soil amendment, the C:N ratio of the raw materials determines their soil effects. Use more nitrogen-rich, green materials for faster decomposition and greater nutrient release. Use low-nitrogen, browner materials to build more organic matter.

The drawback of any type of composting-in-place method is that it requires space and time for the material to decompose. Chopped and mulched residues on the soil surface can also look messy. While adding heaps of organic material to the soil retains water and moisture, it also keeps the soil cooler, which can mean waiting longer in the spring to plant early vegetables. Nonetheless, as your in-place composting system builds along with the health of your living soil, raw materials will transform into new, rich soil right before your eyes. This all happens without the hassle of building compost piles or turning in amendments.

Sheet Composting

Sheet composting, *sheet mulching*, and *lasagna gardening* are all names for adding layers of raw organic materials directly to the soil surface. These methods balance the ratio of green to brown material to build compost in place. Placing this organic food on the surface of the soil attracts worms and other soil organisms from deeper in the soil. As they move vertically, worms incorporate the new compost by tilling the soil. The end result is a soil enriched in nutrients and organic matter, with a structure that has been loosened and improved by biological activity—all without firing up the rototiller or sharpening the digging fork.

For established garden beds, build sheet layers directly on the soil surface, as if you were building a compost pile. I make each layer 2 to 6 inches thick to build a sheet compost that is between 12 and 18 inches in total. Layers include anything that might normally go into the compost pile, including kitchen scraps, manure, chopped green and brown yard debris, coffee grounds, grass clippings, straw, and shredded leaves. As with traditional compost, avoid weeds that have gone to seed or pernicious weeds that spread by rhizomes. Because a sheet compost doesn't heat up like a hot compost pile would, this method will not kill seeds and pathogens. It's more difficult to keep pests out of a

A sheet compost builds garden soils from the ground up. Over the season, messy layers of organic materials, including kitchens scraps and garden waste, transform into rich, loose garden beds.

sheet compost, so definitely avoid any meat or dairy products. For the same reason, keep kitchen scraps buried in deeper layers.

Follow the basic compost-building principle of alternating nitrogen-rich green with carbon-rich brown materials. Finish the topmost layer with bulky mulch material, such as shredded leaves or straw. The surface of this topmost layer will end up drying out, but it is essential to keep the underlying layers moist and actively composting. Over the season, the sheet compost will shrink as it decomposes. Just like tending a compost pile, water the sheet compost to keep it moist and active.

You can also use this method as a way to start a "no-till" garden. We'll cover this as a method of cultivation in Chapter 7. Briefly, this method simply places a smothering layer of cardboard over existing vegetation. Build alternating green and brown sheet compost layers on top of the compost.

The convenience of sheet composting is that you can add layers as they become available. You can, for instance, cover a bed slowly with the scraps from your kitchen, or add heaps of grass clippings at a time as you mow your lawn. Whenever adding a high-quality green material, you'll need to cover it with a brown material to avoid nitrogen losses. Particularly cover kitchen scraps well to prevent pests, messiness, and bad smells.

When using this method, I like to make a thick sheet compost that actively heats up. When I do this, I'll rotate the beds I am actively sheet mulching out of production. I'll

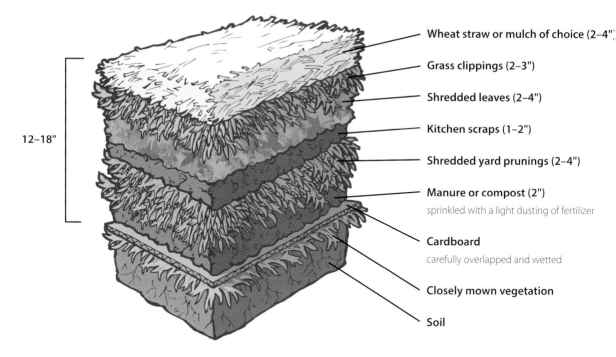

12–18"

Wheat straw or mulch of choice (2–4")

Grass clippings (2–3")

Shredded leaves (2–4")

Kitchen scraps (1–2")

Shredded yard prunings (2–4")

Manure or compost (2")
sprinkled with a light dusting of fertilizer

Cardboard
carefully overlapped and wetted

Closely mown vegetation

Soil

Amending the soil with a sheet mulch is easy. Build up alternating layers of brown and green materials, each 2 to 6 inches thick, for a total sheet-mulch depth of 12 to 18 inches. Build sheet mulches directly on bare soil or smother existing vegetation by using a bottom layer of cardboard.

generally wait until the following spring to plant into them. If the timing is right, however, I might plant crops that tolerate the heat of nitrogen decomposition, such as squash or tomatoes, in the early summer.

If you don't want to take space out of production, add thinner sheet compost layers more slowly, as materials become available. This keeps the compost from actively heating up and harming plants. Layers added in this way function more like a mulch than an active compost pile.

Even added slowly, this way of sheet mulching still amends the soil by adding organic matter while allowing you to use the material at hand in place. When I use this method, I'm thinking about how to minimize my efforts. I'll mulch with the materials closest at hand, nearby garden residues or grass clippings. Once added, I'm sure to keep the mulch adequately watered and covered with a capping brown material.

Another twist on sheet composting is to build the layers on the surface and then till them into the soil. This speeds up the decomposition process so that you can plant sooner. Tilling the sheet compost into the soil, however, disturbs the soil ecosystem. Because of this, it does not take advantage of the highly effective soil-building action of active burrowing organisms.

Trench Composting

Another time- and labor-saving way to amend the soil is with trench composting. Instead of building compost in a pile or in sheet layers on the soil surface, raw organic materials are buried directly in the soil. In this way, the soil itself becomes the compost bin.

Trench composting is all about location. It is like Meals on Wheels for soil organisms, bringing food to where these critters live—in the soil. It also adds plant nutrients right where plants need them—in the root zone. Burying organic materials is also one possible solution for a compacted soil, particularly when light organic materials, like tree leaves, are used. Trench composting also saves time and labor involved with building and transporting compost.

The one disadvantage is that materials take longer to decompose when buried. Depending on the type of material added, you may need to wait months or a year before planting into the trench compost.

Trench composting, because it completely covers the residues with soil, solves the problems associated with smells, pests, or appearances. So if you want a low-work, easy, and smell-proof way to amend your soil and you don't mind waiting a bit, trench composting may be the answer for you. There are several different ways to use trench composting, depending on your goals and how much material you have available.

Digging a hole and burying kitchen scraps is a simple, scaled-down version of trench composting. The year after you bury kitchen wastes, plant directly into the hole to give new plants a nutrient boost.

1. **Dig a hole.** This may be the simplest type of trench composting, particularly if you only have a little bit of material at a time. It is a great way to empty your kitchen compost bucket week after week. Simply dig a hole between 8 and 12 inches deep, add your kitchen scraps, and cover with soil. After one year, plant directly into the hole. You can do this repeatedly to amend an entire garden bed, or save the spot for a specific tree or shrub. The trench compost will give your new plantings a good nutrient kick to get them going.

   If you want to speed up the process, put your kitchen scraps through a blender before adding them to the hole. This is like feeding your soil a vitamin smoothie. If materials are blended before buried, you may be able to plant the hole in a period of months. To see if it's ready, simply dig up part of the hole you filled. If you can recognize the material you put into it, then it needs longer to compost.

2. **Dig a trench.** This is for the composter who has a lot of material at hand. Dig your trench 12 inches deep and 18 inches wide, and fill it with balanced green and brown organic material. Cover the compost with soil. For trench composting to work, you'll want good soil contact with the organic materials that you added. To do this, you'll need to compact the covered trench in some way, either with machinery, a heavy roller, or by hand. Since nitrogen gets tied up as the material composts, it's best to wait a year or a season before planting into the trench.

Trench composting builds compost directly into the garden bed. This can give a boost to a new garden bed or fluff up a compacted soil. The downside is you have to wait a year before planting.

3. **Dig a path.** If you like the benefits of trench composting but don't want to take your garden out of production, you can trench garden paths instead. Trench garden paths to 12 inches deep. Fill with composting material and cover with soil. As you walk over the paths, you will naturally compact the trench compost. The composted pathways can be dug up the following year and placed on top of the existing bed. I like this technique to compost fresh manure. As the manure composts over the season any leached nitrogen will go to feed the neighboring garden bed.

A variation of this is to use paths in an annual garden as part trench compost rotation. Divide your garden into a series of three long strips. In year one, a new trench is filled with raw compost materials. This can happen in a matter of days with stockpiled material, or can be spread out slowly as material becomes available. In year two, the trench is covered and becomes a garden pathway as it composts. In year three, the composted path becomes the new garden bed.

Trench Compost Rotation

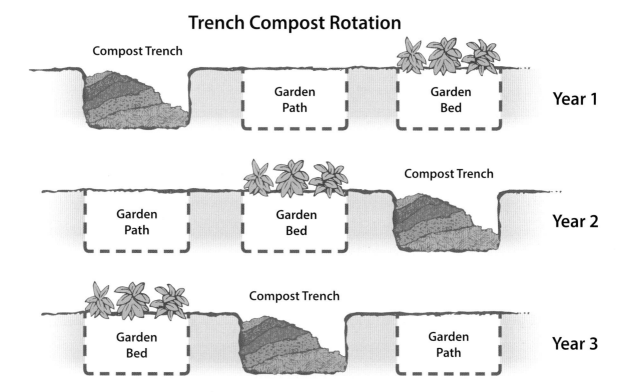

A three-stage trench compost rotation saves garden space, while enriching the soil. In year one, fill the compost trench with raw materials. In year two, cover the trench with wood chips and use it as a path as it composts. In year three, plant your new garden bed directly into last year's path. Continue this rotation to annually refresh garden nutrients.

Lop It and Leave It

"Lop it and leave it" is a method of sheet mulching, but I want to give it special mention. This technique is so deceptively simple that people many times forget that it is one of the easiest ways to amend the soil with organic materials. To lop it and leave it, simply leave organic material where it falls. If you have the time and energy, chop and lop it a little bit to increase decomposition.

In a gardening culture where we've been taught to weed and clean and rake, this can go against the grain. Nonetheless, this is one of the easiest ways for us to build our soil. There are as many examples of how to use this as there are different areas in your gardening. Some common ones that deserve special mention include the following:

1. **Grass-cycling.** Take the bag off your lawnmower and let the grass clippings fall where they may. Contrary to popular belief, leaving grass clippings on the lawn does not lead to dead brown spots if you have a healthy living soil that can decompose residues. Recycling grass clippings on the lawn returns nutrients and organic matter to the soil, leading to healthier and greener pastures.

   As an aside, to maximize the health of your lawn, consider clipping it a little higher than usual, at least ½ to 1 inch in height, depending on the type of lawn grass you have. The higher stubble height of a mown lawn promotes root

Eliminate the work of bagging and hauling, while improving lawn fertility and reducing fertilizer needs. A mulching lawn mower shreds and spreads clipped grass for quick decomposition. You'll never have to bag again!

Chop fall residues directly into the garden bed. As they decompose over the winter, they loosen the soil, while enriching nutrients and organic matter. These French breakfast radishes had sported long ago, so chopping them in was a way to avoid wasting them.

growth, leading to more organic matter production and healthier soils. Shorter stubble heights, on the other hand, starve the roots and stunt growth.

2. **Trees love leaves.** A deciduous forest has some of the richest soils around. Why? Every year, tons of nutrients and organic matter are added to the soil as the trees drop their leaves. Instead of raking, bagging, and hauling, you can simply pull them under existing orchard or landscape trees to enrich the soil in potassium and organic matter. Keep the leaf mulch 2 to 3 inches away from tree trunks to prevent rot.

3. **Weeds are feasts.** Get rid of weeds and amend your soil at the same time. Instead of raking and bagging weed debris, let them lay where they fall when mown with a weed whacker. Leave them as mulch or dig them lightly into the soil. If you're using weeds in this way, just make sure to get them before they go to seed. Also beware of pernicious nasty weeds that reproduce readily from roots or even cut stems. Chapter 6 highlights some of these particularly problematic pests.

4. **The fall chop.** A messy garden is a productive garden. At the end of the season, don't spend those lovely autumn days ripping up all of last year's bounty. Let it compost in place to enrich your soil for the next year. Start by removing particularly woody stems. Then take a sharp-bladed spade, hoe, or other implement of choice and chop away. Use the chop to split garden debris into 3- to 6-inch pieces and partially bury them in the soil surface. If you have other fall amendments to add, do it right on top of the debris and chop or till the whole package in together.

Compost

Compost is renowned for its effectiveness as a soil builder. It has become the staple amendment for many organic farms, backyard gardens, and perennial landscapes. At the same time, it provides a simple and easy way to recycle and reuse organic materials that range from kitchen scraps to yard debris. On a large scale, it also provides a way to recycle lumber and agricultural byproducts.

Compost is made by adding raw organic materials to a pile, bin, or windrow. The microbes consume the raw organic materials and transform them into rich, dark organic matter. This is the same process of decomposition that occurs in the soil itself. In a compost pile, however, we can control the process and the final product that we add to our garden soils.

During composting, roughly half of the original volume is lost to decomposition. The quality (C:N ratio) of the final product depends on the original quality of the raw materials. Generally, finished composts are of intermediate quality, with a C:N ratio from 20:1 to 30:1. This means that compost has a relatively low nutrient-supplying power. You'll likely need to supplement with another nutrient-rich source.

In some ways, composting is like pre-chewing the food that we feed to the microbes. It creates a more stable amendment with the longer-lasting effect of building organic matter. Although compost doesn't add many nutrients, it also won't tie up nutrients as it decomposes. Because of this, we can plant into compost-amended soils immediately.

As a soil-building amendment, compost is unparalleled. It is safer to use than manures, which can concentrate salts or harbor pests. High-temperature composting kills weed seeds and pathogens. Compost bins also save garden space. Sheet composting requires time and space for materials to break down, and green manures need dedicated garden soil. Making compost, on the other hand, requires only a corner of your yard. If turned frequently, this dedicated space can continually turn out rich soil-building material. Adding compost to perennial or ornamental gardens has a clean and neat look when used as mulch. Compost takes up less space and is easier to truck than raw materials. Healthy compost, as a breeding and feeding ground for microorganisms, also provides the important and often overlooked benefit of inoculating the soil when spread around the garden.

Compost has its disadvantages as well. Purchased compost can be quite costly. Both homemade and purchased composts vary widely in quality, and the low nutrient content of compost may not meet your soil's fertilization needs. Building good-quality compost requires a lot of sweat equity: building and turning the piles, carting compost around the garden, and forking it into the garden soil.

The creative composter, however, can find a wealth of free organic materials in neighborhoods and communities. Kitchen scraps from school cafeterias or restaurants, leaves from city collection, grass clippings from neighbors, spent grain from breweries, and brush trimmings from the local arborist can all be turned into rich, black garden soil. This, in turn, is transformed into food and flowers that are shared with your local community. In this way, making and using compost extends the boundaries of your whole-garden ecosystem beyond the edges of your own yard.

Using Compost

For the dedicated home gardener, compost is an indispensable tool. Use compost supplemented with fertilizers to add organic matter to soils when preparing new gardens or lawns. For new lawns, spread 1 to 2 inches of compost on the soil surface and incorporate to a 6- to 8-inch depth. If tilling compost to more shallow depths of 3 to 4 inches, use half this amount. To prepare annual gardens, incorporate compost at the rate of 2 to 3 inches to a 6- to 8-inch depth in the fall or spring before planting. The effects and nutrient-supplying power of compost are cumulative. In the fourth year of annual compost applications, reduce the rate to 1 to 2 inches of compost, incorporated to a 6- to 8-inch depth.

Compost Basics

We've talked about the needs of the living soil: food, water, shelter, and air. The living compost pile has the same requirements. The compost pile needs a mixture of green (high-nitrogen) and brown (low-nitrogen) food for the soil microbes. You want to add these in a rough ratio of one part green to two parts brown material, for an overall C:N ratio of 30:1. This ratio will generally provide the nutrients microbes need, while keeping the pile aerated and well structured. If green material is in short supply, then a nitrogen fertilizer, such as bloodmeal or fish emulsion, can be added instead. Some types of materials, such as horse manure mixed with bedding or chipped shrub trimmings, inherently have the right C:N ratio and can be used as-is.

A compost pile needs enough water so that the microbes can be active, but not so much that the pile becomes waterlogged. The pile should feel like a wrung-out sponge,

THE RAW STUFF: GREEN, BROWN, AND GREENISH BROWN

A balanced compost pile needs a mixture of nitrogen-rich green material and carbon-rich brown material to feed the compost organisms. All-in-one materials inherently have the right mix of carbon and nitrogen. These all-in-one self-composters also do well in the layered compost pile or sheet mulch, although they don't need other materials to balance them out.

Green Materials (Nitrogen-Rich)	Brown Materials (Carbon-Rich)	All-in-One (Green + Brown) Materials
Fresh grass clippings	Chipped woody yard waste	Chipped green tree or shrub trimming with foliage
Kitchen scraps	Wheat straw	Shredded fresh deciduous leaves
Alfalfa hay	Sawdust	Mature garden residues
Fresh manure	Cardboard	Aged hay or straw
Fresh garden residues	Dried leaves	Horse manure with bedding

moist, but not dripping water if squeezed. Protect the compost pile from drying out in the sun or getting drenched in the rain by covering it with a tarp and locating your pile in a shady spot. Water your compost pile in the summer to keep it from drying out.

If the pile does become waterlogged, piles turn anaerobic. When this happens, microbes suffer and it takes a long time for the pile to break down. Meanwhile, keep your neighbors at a distance. An anaerobic pile smells like rotten socks dipped in sour milk. This is not exactly appetizing. A well-aerated pile at proper moisture, on the other hand, should only have a faint ammonia smell as the pile is composting. When the compost is finished, it won't smell at all.

Temperature is another important factor. If too cold, compost bacteria won't be active. Because good compost needs to retain heat and stay moist, there is a minimum pile size that works best for compost-making. Generally, you want to build a pile that is about 1 cubic yard: 3 feet wide by 3 feet deep by 3 feet long.

You may or may not choose to use a compost bin to contain the pile. The advantage of a bin is that it keeps material neat and tidy. If you have pest problems, using hardware cloth or chicken wire (depending on the size of the pest) can keep them out of the compost. The disadvantage of a bin over an open pile is that compost can be harder to turn and to use.

Truly hot compost can eliminate pests, pathogens, and weed seeds. On a home scale, it is difficult to keep piles at the optimal temperatures. For this reason, it is better not to add weed seedheads or rhizomes of invasive plants to compost. Meat scraps, though a great source of nitrogen, are also better avoided in the home compost. They require high heat to kill pathogens and are a big attraction to unwanted animals and pests. If possible, don't use yard debris and grass clippings that have been sprayed with herbicides, as these concentrate in the compost. Keep this in mind when collecting green materials from your neighbors.

Hot Composting

Hot composting is the most time-intensive composting method, but it pays off by producing a good-quality product in a relatively short period of time. If done properly, this method also kills weed seeds and manure pathogens. The goal in hot composting is to heat the pile to 120 to 150° Fahrenheit in several rotations. To kill weed seeds, the pile needs to reach 145° Fahrenheit. For the pile to heat properly, it must have a minimum size of 3 by 3 by 3 feet, have the proper ratio of green to brown material, and be kept at the correct moisture conditions.

continued on page 106

THINGS TO AVOID IN THE HOME COMPOST
The sky is (almost) the limit when it comes to what you can add to a compost pile. The following table gives some of some things to avoid in the home compost.
Weeds that spread by rhizomes
Seedheads
Cedars
Herbicides
Unshredded leaves (shredded leaves are great!)
Meat scraps and animal byproducts
Human and pet manure

Use a compost thermometer (or your bare hands) to monitor compost temperatures. To kill seedheads and pathogens the ideal temperature is 145° Fahrenheit. Once temperatures start to decline, it's time to turn the pile.

THE PERFECT PILE

Follow these steps to build hot compost that will be ready in two to three months.

1. Collect enough "brown," carbon-rich material and fresh, "green," nitrogen-rich material to build a 3-by-3-by-3-foot pile.
2. Find a shady, level spot for your compost pile.
3. Remove any thatch, vegetation, or stones from the footprint of your pile, roughly 2 feet wider than the size of your pile. Level the ground and loosen the top 1 to 2 inches of soil with a fork.
4. If you are concerned about pests, you can use a pest-protected compost bin, either bought or made at home using hardware cloth. You can also build compost without a bin to hold it. Whether you use a bin or not, compost built directly on the ground, with an open bottom, encourages soil fauna to come up into the compost.
5. Place some material at the bottom of the pile to help with aeration. Wood chips, stalks, or brushy sticks are useful for this purpose.

Make a homemade, pest-proof compost bin with hardware cloth and lumber (top photo). If you're regularly turning hot compost, using a bin adds extra work. For slow composting, a compost bin conveniently holds material as you add it slowly over the months. There are many types of commercial (bottom photo) and homemade bins to choose from. See the Additional Resources section for information on how to build your own.

6. Chip, grind, or chop your raw compost materials into small pieces. You can use a lawn mower to shred leaves, a chipper for woody debris and tree trimmings, and a sharp spade for garden debris and kitchen scraps. Small pieces give more surface area for decomposition.
7. Begin building your pile. For a 3-by-3-foot open pile, start the base of your pile at 4x4 feet, as material tends to slope upward. Layer one part green to two parts brown material. Mix the layer with a pitchfork and add water until it is moist, like a wrung-out sponge. Continue adding material in layers, mixing and watering each layer as you go. Try to keep the sides vertical as you build the pile upward.
8. Add a topmost layer of carbon-rich brown material.
9. Cover the pile with a tarp or compost bin cover.
10. Check the center of the pile periodically to see if it generates heat. If not, you may need to adjust moisture conditions or add more nitrogen to the pile. When the pile has cooled down (after about one to two weeks) turn the pile over by mixing the outside of the pile into the center and the center to the outside. Try to keep the dimensions of the pile roughly the same.
11. Periodically check your piles to see if they need water.
12. When the compost no longer heats with turning; does not have an odor; and looks relatively uniform, brown, and crumbly, it is ready for garden use.

Turn a hot compost pile by mixing material from the outside of the pile to the center and the center to the outside. Add water as necessary to keep the pile evenly moist, but not wet.

continued from page 103

After the first time the pile heats and cools, it is necessary to turn the pile every few weeks. Once the compost does not heat with turning, it is ready for use in the garden. Generally, finished compost takes several months, though if turned more frequently, it can take as little as a month. The disadvantage is that turning compost is labor intensive. It also requires having a good amount of raw materials on hand, including stockpiled brown and fresh green material for the initial build.

Slow Composting

Slow composting is a good option for the gardener who still wants to recycle kitchen and yard wastes but doesn't have the time or resources to continually turn a pile. Slow composting is also known as passive or cool composting, because these piles never reach the high temperatures of hot compost. The advantage of slow composting is that it requires much less work. Simply pile up the material, sit back, and let nature take its course. The disadvantage is that the compost takes much longer to finish—from six months to a year.

Use the same pile-building principles that apply to hot compost. The difference is that you can slowly build your pile as material becomes available. As you layer material, keep in mind that you'll still want to have a roughly 1:2 green-to-brown ratio for a good quality compost. The best way to do this is to keep carbon-rich brown materials on hand. As you add green yard wastes or kitchen scraps, add some of the brown material to balance the ratio.

If you have the time or the incentive, give the pile a turn every once in a while. Though not creating the hot temperatures of fast compost, this will help speed the process and create a more uniform product. You'll also still want to keep the compost covered and add water from time to time to keep good moisture conditions. When your pile reaches that magic 3-by-3-by-3-foot size, start a new pile and wait for the old one to finish. You can also harvest finished compost from the center and bottom of the pile and continue adding new material to the old pile.

Composting with Worms

Composting with worms creates a valuable, nutrient-dense soil amendment by recycling household waste. This method, also known as vermiculture, feeds kitchen scraps to a bin stocked with red worms. The result is a rich compost of worm castings that is ready in as little as three months.

Worm compost, which is essentially worm manure, concentrates nutrients. Although producing less bulk material than traditional compost, the final product is composed of rich, nutrient-dense packages. High in nitrogen, phosphorus, and potassium, with a C:N

Harvested worm castings are power-packed nutrient pellets, highly concentrated in nitrogen, phosphorus, and potassium. The perfect round, soil aggregates, they improve soil structure while fertilizing the garden.

ratio of 15:1, worm castings supply readily available nutrients to plants. Worm castings also work wonders for stimulating the living soil, both by adding food and by inoculating with living organisms.

Composting with worms is a high-efficiency, low-labor option for the small-scale composter who mainly wants to recycle kitchen scraps. Add high-nitrogen kitchen wastes continually to the worm bin, instead of building a compost pile all at once. The worms do the work of turning the compost for you, speeding the compost process. The main disadvantage of this method is that a single bin yields a relatively small amount of castings at one time, so it may not effectively cover a large garden. However, since you can harvest castings every few months, you'll have a continual supply of material to spread around. Make the most of this valuable material by incorporating it into the soil before planting a perennial tree or shrub, adding it to potting mixtures, or fertilizing vegetable and flower beds. It's also a great ingredient for compost tea.

The basic needs for your worm compost are the same for traditional compost: food, water, shelter, and air. Vermiculture literally means "cultivating worms." You'll have the best success with your worm bin if you think of composting as farming worms, instead

The red wrigglers that stock your worm bin transform kitchen waste into a valuable, nutrient-dense amendment. These little creatures are massive consumers, eating from 25 to 35 percent their body weight every day.

of building compost. Take care of the worms' needs by not letting the bin get too hot or cold, or too wet or dry, and by not overfeeding or underfeeding your worms.

Worm bins use a specific species of red wigglers, *Eisena fetida,* available online or at certain garden centers. Eventually you'll want a pound of worms for every cubic foot of bin space. Because the worms reproduce quickly, you can start with buying a single pound of worms that you feed slowly for the first month as they grow in numbers.

To become a worm rancher, your first consideration is bin location. Red wigglers need temperatures of 65 to 80° Fahrenheit—in other words, they don't like it too hot or too cold. In some climates, you'll need to insulate your bins in the winter or keep them indoors to keep the worms alive. If you want your worms to live outside, find a shady and sheltered spot. I like to place my bin under the eaves of the north side of my house.

You can buy a worm bin or use the DIY approach. A plastic tub is simple to make and to transport from inside to outside. A wooden bin, made with untreated lumber, provides better overall airflow. In either case, the worms, like any other soil critter, need air and water. That means the bin needs good drainage and insulation. Because red wrigglers don't like the light, a tight lid is also essential.

A homemade wooden worm bin can be kept neat and tidy in a basement or garage. Good ventilation, drainage, and a tight lid are essential. Use a pan to collect the water that drains from the bin. This water is also a valuable source of nutrients to spread on garden soil or compost piles.

Size the footprint of your bin based on the food you'll be adding. You'll want roughly 1 square foot for every pound of food scraps you produce in a week. For a family of four, this may come to about 5 to 7 pounds per week. This translates to a roughly 2-by3-foot bin. Your bin can be from 8 to 18 inches in depth. Drill ¼- to 1-inch drainage holes along the sides and on the bottom. Indoors, use a pan to collect drainage water, but make sure this doesn't keep the bin from draining. Use this nutritious drainage water to feed plants or as a compost tea ingredient. You'll find sources for more detailed worm bin plans and designs in the Additional Resources section.

Add food and bedding to the worm bin at roughly a 30:1 ratio of bedding material to food. Bedding material supplies the carbon for the composting process. It also keeps the bin aerated and dry. Use a combination of several types of bedding material to keep worms happy. Examples of readily available bedding material include paper, paper bags, cardboard, straw, sawdust, leaves, and dried-out grass clippings. It's good to have at least some coarse material in your bedding to prevent it from matting and restricting airflow.

Worm food is any of the nitrogen-rich food scraps that come out of your kitchen and the rotten fruits and vegetables from your garden. Nitrogen-rich coffee grounds are also great additions. Meat scraps, dairy, oily foods, and cooked food high in salt are not good for the worms. Oily foods tend to make the bins too acidic and slow down the compost process.

Because worms don't have teeth, shred, chop, or grind bedding and food into small pieces before you add them to the bin. You'll also need to add a bit of sand or grit, which helps the worms break up the food. When you first start feeding your worms, you'll want to take it slow. Work your way up to the target feeding level over the course of a month.

Once you've got your bin in place and your worms in hand, you'll want to take the following steps to build, maintain, and harvest your worm compost

Starting Your Bin

1. Fill the bin three-quarters full with moist bedding. The bedding should be just lightly dampened to the moisture level of a wrung-out sponge.
2. Mix in a few handfuls of sand or garden soil to the bedding material.
3. Pull back the top layer of bedding and add a layer of food. Replace the bedding.
4. Add the worms to the top of the bedding. They should quickly disappear into it. Make sure the bedding is well fluffed to give the worms plenty of space to breathe.

Care and Feeding

1. Check the box regularly for food and moisture. If a foul smell starts to form, you may be overfeeding. Try adding food more slowly, breaking up the feed more, or adding more bedding. Smell might also indicate problems with drainage or ventilation. Make sure bedding is not matted and drainage holes are not blocked.
2. Continue to feed your worms once a week, placing food in a different spot each time. It is best to collect food in a sealed container to prevent flies. Find the amount and timing of feeding that is right for your bins. Bedding will be eaten more slowly. Once it starts to disappear, replace with moistened bedding material during feeding, careful to keep the 30:1 ratio of bedding to food.
3. Continue to monitor temperature, moisture, and smell and make adjustments as necessary. In the winter, it may be necessary to move the bin indoors or insulate it with straw or leaves. Worms will not survive a freeze.

When castings begin to collect at the bottom of the bin, after about three to four months, it is time to harvest your vermicompost. This involves separating the castings from the worms and the unused bedding. Shining a light helps drive the worms into the open for collection. Once the worms are collected, weigh them to find out how many you need. If you have more than 1 pound per cubic foot, you can harvest the extra to give to friends, start a new bin, or add to your garden. Use the leftover bedding to start the new bin. There are several harvesting methods:

Feed worms weekly by pulling back the bedding and adding a thin layer of kitchen scraps. Add food to a different spot of the bin every time.

1. **Hand sorting.** With this method, you dump the bin on a tarp or plastic sheet and sort everything by hand. You'll end up with clumps of castings and bedding and a pile of worms. Keep your worms happy by giving them a mound to burrow in as you sort.
2. **Screening.** Empty the bin onto a screen by ½-inch hardware cloth tacked to a wooden frame. Empty the bin onto the screen, collecting the castings below and the worms on top of the screen. Collect the unused bedding for your new bin.
3. **Divide and conquer.** Divide the bin into two sides by pushing all of the contents of the bin to one side and providing fresh bedding and food on the other side. Continue feeding on the fresh side only. After two to four weeks the worms will move into the fresh material, and you can harvest the finished compost from the other side without much sorting.

Use the shredder function on your electric blower to collect leaves in the fall. Empty the collection bag into your compost bin or spread the shredded leaves directly on your lawn or garden beds.

Leaf Mold

Composting deciduous leaves deserves special mention. The leaves from hardwood trees like oak, elm, maple, and other deciduous species are good bulking brown materials for a compost pile or worm bin bedding. They are also a wonderful mulch by themselves.

Because leaves tend to form mats that won't let water through, it's a good idea to shred them before use. Shred leaves by spreading them out and running over them with the lawn mower, placing them in a large garbage bin and using a weed whacker, or purchasing a leaf shredder specifically for this purpose.

Composting leaves by themselves results in a superior soil amendment called *leaf mold*. Leaf mold is unparalleled as a soil conditioner, particularly for heavy clays. It fluffs up a compacted soil, improves overall structure, and dramatically increases water-holding capability. At the same time, leaf mold adds nutrients and builds organic matter. When fresh, leaves have a C:N ratio of about 50:1. As they compost, they become more nutrient rich for a final C:N ratio of about 20:1, which adds some nutrients to the soil. Leaf mold is also unique among soil amendments in that it has a near-neutral pH and is naturally rich in potassium.

Making leaf mold is incredibly simple. Collect as many fall leaves as you can from your home and your neighbors. In cities with curbside leaf collection, it is sometimes possible to get a truckload delivered to your home. To make leaf mold, first shred the leaves to make smaller pieces and increase the speed of composting. Pile the leaves as you would normal compost, with a minimum pile size of 3 by 3 by 3 feet, slightly moistening the material as you go. Let the pile sit in a corner of your yard, keeping it covered and adding moisture as necessary.

Collect fall leaves into a bin, a plastic bag, or a pile. Let them rot over the winter to create leaf mold, a low-cost, low-labor amendment that does wonders to improve soil condition.

In about six months to a year, you should find decomposed, fluffy brown material that you can then add to your garden or mulch around perennials. If you've collected your leaves in the fall, this means that you can probably incorporate the leaf mold for spring planting, depending on winter temperatures in your area. To increase the speed of composting, turn the pile a time or two over the winter. Because leaves are free and plentiful in many areas, this is a low-cost, low-labor method to produce an amendment that the living soil will love.

Manures

For many gardeners, manure is low-cost and plentiful soil-building resource. Forget Texas crude oil; manure is the real black gold. Manures are concentrated in a broad spectrum of essential macro- and micronutrients, including nitrogen, phosphorus, potassium, and sulfur. They also build long-term levels of soil organic matter. Over time, this increases the nutrient-supplying power of the soil to replace fertilizers altogether.

Depending on where you live, you may be surrounded by people wanting to get rid of this garden treasure. Some ranchers or hobby farmers will sell it by the truckload, while others will let you take it for free if you load it yourself. Bagged, composted manure is also available at garden centers. These products tend to be higher in salts and lower in nutrients. Sometimes, bagged poultry manure is not completely composted. If it gives off an ammonia smell, let it mellow, either in the soil or a pile, before planting into it.

Nutrients in manures vary significantly depending on the type of animal, whether and how much bedding was added (for instance, straw or wood chips), and whether it is composted or fresh. In fact, fresh manures can have C:N ratios ranging from 5:1 to 30:1. Generally, fresh manures without bedding will release nutrients, although it is difficult to know the exact amount of nitrogen, phosphorus, and potassium available without having the manure tested. Chapter 5 provides a table of average manure nutrient contents as a starting point. From this you can estimate the nutrients supplied by manure applications.

Composted manures provide fewer nutrients, with C:N ratios of 20:1 to 30:1. On the other hand, they are highly effective at building soil organic matter and fill up soil nutrient reservoirs in the long term. With composted manures, you may need to add supplemental fertilizers in the first year or two of application. This is particularly true if the manure contains a lot of carbon-rich bedding material, which may tie up nitrogen temporarily.

Over time, the nutrient supply from both composted and fresh manures is cumulative. That means that the soil-building effects keep increasing. Nutrients are released

Prevent pollution, pathogen problems, and nutrient loss by storing manures properly. Keep manures well-covered from rain to prevent leaching. Ideally, store manures on an impervious surface, like a concrete driveway.

not only from the current year's manure application but from manure applications from the previous one to five years. As these nutrient sources build up in the soil, additional fertilizers become unnecessary.

Just remember that too much of a good thing can be bad. Manures are high in salts and can oversupply certain nutrients, so don't apply too much in a single season. Monitor nutrients and salts with soil tests, and take a break from manures altogether every once in a while. After about five years of building soils with manures, use a different amendment. Soil testing can guide you in this regard. If you live in an area prone to salt buildup in soils, you'd be wise to avoid manure altogether. Have questions about the manure's effects? Have the manure tested (see the amendment quality guide earlier in this chapter).

To use composted manure, spread 1 inch evenly over garden soils and incorporate to a depth of 6 to 8 inches. Add nitrogen-rich fresh manures more sparingly, from ½ to ¼ inch. Don't lose nitrogen to the air. Incorporate manures within twelve hours of spreading.

Using Manures

Fresh manures can replace fertilizers altogether. To supply the most nutrients, use them in the spring before planting. Apply and incorporate fresh manures about a month before starting your spring garden to give the manure time to mellow. This is particularly important when you're using fresh poultry manure. It's high in ammonia and can "burn" plants if applied directly.

Fresh manures, however, can harbor nasty pathogens like *E. coli*. To minimize food-safety risks, as opposed to maximizing nutrients, fresh manures are best applied in the fall. This gives pathogens time to break down in the soil. In fact, organic agriculture standards require that fresh manures be applied four months prior to crop harvest. Unfortunately, this reduces the nutrients available to plants during the growing season. Find what works for you to reduce risk and maximize nutrients.

It is also critical to handle and store manures safely. Don't let fresh manures touch the parts of the plants you plan to harvest. This is particularly important for leafy greens and fruits eaten raw. Wash your harvest thoroughly before eating or distributing the food. Of course, wash your hands and tools after handling manures to avoid cross-contamination. Hot composting at temperatures of 145° Fahrenheit can also kill most pathogens. To ensure manure safety, follow the instructions for hot composting, and make sure your pile reaches the correct temperatures.

It is also important to store manure carefully. Make sure that it is covered properly. Letting rainwater run off or leach your manure pile not only removes valuable nitrogen, it can pollute surface and ground water. Water leached from compost can also contaminate other parts of your garden with pathogens.

CHICKEN TRACTOR

One of the simplest and most effective ways to apply manure is to let the animals do it for you. If you're lucky enough to have backyard chickens, use them to fertilize your garden. Use a chicken tractor, consisting of a mesh-covered portable frame, to move chickens around your lawn and garden. Not only is the chicken poop good for your soil, it keeps the chickens clean and provides a variety of fresh feed.

To fertilize a lawn, move the chicken tractor daily. This adds nutrients without damaging the grass. Move your chickens into garden beds at the end of the season. They'll add manure, clean up residues, reduce insect pests, and lightly till the soil. Because chicken manure can still harbor pathogens, it's a good idea to wait several months before planting into freshly fertilized soil.

Backyard chickens are an increasingly common feature of home gardens. Integrate chickens into your soil-building program by using a chicken tractor to till and fertilize the soil.

Because fresh manures are nutrient dense, limit applications to less than ½ inch. Spread the manure on the soil surface and incorporate 6 to 8 inches deep. For high-nitrogen poultry manures, apply less than ¼ inch. When exposed to the atmosphere, nitrogen is quickly lost through volatilization. Because of this, be sure to incorporate fresh manures within twelve hours of spreading.

Composting manures reduces risk from pathogens. To amend garden soils prior to planting or to prepare a new lawn or perennial bed spread 1 inch of composted manure and mix to a depth of 6 to 8 inches. Apply ½ inch if tilling to only 3 to 4 inches or if top-dressing in a perennial garden. You can also mix composted manure into planting holes for shrubs and trees or use as mulch around perennials.

Manure is also a great green material for both pile and sheet compost. Be sure to add some brown material to balance out nitrogen-rich hot manures. Manures containing bedding often have the right green-to-brown ratio to compost alone. Horse manure, for instance, has a C:N ratio of 30:1 and doesn't require any green or brown additions.

CHAPTER 5

A BALANCED DIET

FERTILIZERS AND WHOLE-SOIL FERTILITY

There is no doubt that a living soil, amended regularly with organic material, can meet soil nutrient needs. Nonetheless, as you start the process of building healthy soils, organic matter alone may fall short of plants' nutrient needs. When this happens, you can use concentrated fertilizers as a temporary fix to fill in the gaps.

This chapter fills in the blanks between what nutrients a soil has and what nutrients a soil needs. I like to call this approach *whole-soil fertility*. With this method, fertilizers are added as supplements rather than standards. Whole-soil fertility accounts for all the different nutrient sources in your soil. This includes the nutrients released by the living soil and organic matter over the growing season, as well as the nutrients measured by a soil test. In this way, we use the minimal amount of concentrated fertilizer necessary to meet plant needs. Instead, we rely more and more on the living soil and the nutrients recycled in a sustainable whole-garden system.

The following pages will give you the tools needed to estimate the nutrients provided by the living soil and create a whole-soil fertility plan. A step-by-step worksheet determines exactly what you'll need to add to meet plant demand for nitrogen, phosphorus, and potassium. We'll also explore the extensive variety of fertilizers available for today's gardener. In this way, fertilizers become part of a whole-soil-building program fertility plan for the vegetable garden, lawn, orchard, or perennial landscape.

FERTILIZERS

In this day and age there are a range of fertilizers to choose from. The aisles of the local garden center are full of bagged choices. There are chemical fertilizers with various NPK combinations for specific nutrient needs. There are plant and animal by-products, such as soy, alfalfa, blood, feather, and bonemeal. There are natural rock powders and ground-up minerals, like greensand and rock phosphate. You can find bagged manures and composts, liquid fish emulsions, microbial inoculants, and dried seaweeds.

Bloodmeal, feathermeal, fishmeal, and soybean meal are just a few of the certified-organic nitrogen fertilizer choices available.

The sheer number of fertilizer choices can be overwhelming. In the end it comes down to matching soil needs with fertilizer supply. In my opinion, how a fertilizer is used is generally more important than what fertilizer is used. Using fertilizers correctly,

FERTILIZERS AND AMENDMENTS: WHAT'S THE DIFFERENCE?

The difference between a fertilizer and an amendment is a legal one. Technically, a fertilizer is an amendment that, by law, contains a guaranteed minimum percentage of nitrogen, phosphorus, and potassium. This means that when you add 100 pounds of fertilizer, you know exactly how much nitrogen, phosphorus, and potassium you've supplied to your garden. Knowing this allows us to use the right fertilizers to match soil-nutrient needs.

By this definition, all fertilizers are a type of amendment, but not all amendments are fertilizers. The nutrient value of manures and composts, for instance, varies widely depending on many factors, including animal diet and bedding material for manures, and the source material and composting method for composts. The way that amendments are handled and stored can also dramatically affect the fertility of these materials. Even though bulky organic amendments do provide some, and sometimes all, of our nutrient needs, they cannot be sold as fertilizers because the exact nutrient concentration is unknown.

Some commercial suppliers do sell bagged manure products as fertilizers, providing a labeled NPK percentage. In these cases, the manures are analyzed for nutrient content, and supplemental materials are often added to increase the nutrient content to a minimum value.

and with whole-soil fertility in mind, improves the living soil while preventing waste and pollution. As whole-soil fertility builds, fertilizer needs and costs decrease. The choice of which fertilizer to use comes down to several considerations that affect how it will be used:

1. **Organic or nonorganic?**
2. **Nutrient concentration?**
3. **Release time?**
4. **Cost?**

Fertilizer Sources: Organic or Non-Organic

For many gardeners, the question of using organic versus nonorganic fertilizers is an ethical choice. Organic fertilizers come from natural sources, while nonorganic fertilizers are synthetic chemicals. Most people assume that organic fertilizers are more environmentally friendly. While this is often the case, it is not always true. The most environmentally friendly option is to minimize fertilizer use altogether by building living, sustainable soils.

Both organic and synthetic fertilizers, as well as organic amendments such as manure and compost, can cause water pollution. This happens due to improper storage, excessive application, or the wrong time of application. Properly used, organic fertilizers are generally less likely to cause water pollution than synthetic chemical ones. The main environmental benefit of organic fertilizers is that they are often made of recycled plant and animal byproducts, while non-organic fertilizers are made in energy-intensive factories. Nonetheless, some organic fertilizers and amendments, such as Chilean nitrate or peat moss, are mined from nonrenewable resources. Today there are enough alternative fertilizer and amendment options that it is easy to avoid using these unsustainable options.

Other differences between organic and nonorganic fertilizers affect how each is used. Generally speaking, organic fertilizers have lower nutrient concentrations, longer release times, and longer-lasting effects than chemical fertilizers. These factors are discussed in more detail in the following sections.

From a whole-soil perspective, I prefer organic fertilizers because of their longer-lasting and soil-building effects. If using synthetic chemical fertilizers, I advocate applying them with plenty of bulky organic material. Low-quality (low C:N) organic amendments hold on to fertilizer nutrients to provide more sustained, long-term effects.

Nutrient Concentration

From the perspective of whole-soil fertility, fertilizers are used only when there is a nutrient deficiency that is not met by regular organic amendments. For this reason, choose the right fertilizer to match nutrient supply with demand.

Some fertilizers are high in a single nutrient, while others provide a balanced dose of the big three: nitrogen, phosphorus, and potassium, otherwise known as NPK. Organic fertilizers, additionally, often contain a broad spectrum of secondary nutrients and micronutrients in addition to NPK. The tables in this chapter provide nutrient concentrations for some common fertilizer choices. Use these with the whole-soil fertility worksheet at the end of the chapter to match fertilizer choices with soil needs.

Release Time

Release time refers to when and for how long fertilizer nutrients are made available to plants. Knowing release time is essential for matching the timing of nutrient supply with nutrient demand. Organic fertilizers generally have slower release times than nonorganic fertilizers, but there are variations among both types. Release time is also an important consideration for nutrients when using organic amendments.

Most synthetic chemical fertilizers, unless treated with a slow-release coating, dissolve in water and are immediately available. Microbes release nutrients from organic fertilizers by decomposition. In other words, the microbes have to chew up the organic fertilizer and spit out the nutrients. This can happen immediately or can take up to four months, depending on the C:N ratio of the fertilizer. The size of the fertilizer grains also affects release time. Powdered forms are more quickly available than pelleted forms. Because microorganisms require temperatures above 50° Fahrenheit and good moisture conditions to be active, count release time in weeks when soil temperature is above 50°.

Nutrients from organic fertilizers with moderate release times become available about a month after application. Within four months about 75 percent of nutrients are released. The remaining nutrients become available over a period of months to years, benefitting a soil's store of long-term nutrients. Fertilizers with slow and very slow release times take four months or more for nutrients to become available. The bulk of nutrients from these fertilizers are released over the course of years, which builds soil fertility in the long term. Organic fertilizers with fast release times are immediately available and are completely released after about a month. Sometimes, these come in powdered, water-soluble forms that are applied to plant leaves for a quick nutrient burst.

continued on page 126

HOW TO READ A FERTILIZER LABEL

Those dashed numbers on the front of a fertilizer bag aren't some secret code that only farmers and crop specialists can understand. They represent the guaranteed amounts of nitrogen, phosphorus, and potassium (NPK) in a fertilizer bag. For nitrogen, this is the percentage of total nitrogen. For phosphorus and potassium, this is the percentage of phosphate (phosphorus in the form of P_2O_5), and potash (potassium in the form of K_2O) that is available in a single growing season.

Phosphate and potash are slightly antiquated terms still used for fertilizers.

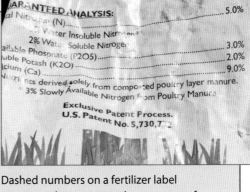

Dashed numbers on a fertilizer label represent the guaranteed percentage of NPK—nitrogen, phosphate (phosphorus), and potash (potassium)—in a fertilizer bag. Use this to determine how much fertilizer to add to meet specific nutrient needs.

Fortunately, plant nutrient recommendations on soil test reports are often made on the basis of phosphate and potash. If it is necessary to convert total P to phosphate or total K to potash, then use the following formulas:

Phosphate $(P_2O_5) = P \times 2.3$
Potash $(K_2O) = K \times 1.2$

It's simple to put these numbers into real-world terms. Let's look at a bag of soybean meal with an NPK (total nitrogen, phosphate, and potassium) of 7–2–1. For every 100 pounds of fertilizer, there are 7 pounds of nitrogen, 2 pounds of phosphate, and 1 pound of potash. For every 50 pounds of fertilizer, there are 3.5 pounds of nitrogen, 1 pound of phosphate, and 0.5 pounds of potash, and so on.

Here's a breakdown of how we figured this out for the nitrogen:

- The label says 7–2–1 for NPK. This means there is 7 percent nitrogen in the bag.
- 7% = 7 ÷ 100 = 0.07
- 0.07 × 50-lb. bag = 3.5 lb. of nitrogen in a 100-lb. bag of soybean meal

If we want to add less than an entire bag of fertilizer, we work backward. Knowing that we want to add 1 pound of nitrogen to a given area of the garden, we use the following formulas:

- 1 lb. (N needed) ÷ 7% (percentage in soybean meal)
- 7% = 0.07
- 1 lb. (N needed) ÷ 0.07 (percentage in soybean meal)
- 1 ÷ 0.07 = 14 lb.

We need to add 14 pounds of soybean meal to get 1 pound of nitrogen in our garden. This is a little over a third of a 50-pound bag of fertilizer.

ORGANIC FERTILIZER CHOICES

Fertilizer choice depends on nutrient concentration, release time, method of application, and other desired benefits. The table below provides a partial list of available organic fertilizers and their average characteristics.

Fertilizer Name	Average NPK *	Release Time**	Form	Notes
Alfalfa meal/pellets	2–1–2	Medium	Solid—plant byproduct	Builds organic matter. Micronutrient source.
Bat guano	10–3–1	Slow	Solid—animal (mined)	Can also get a type processed for phosphorus.
Bloodmeal	12–0–0	Medium	Solid—animal byproduct	One of highest organic nitrogen sources.
Bonemeal	3–15–0	Medium	Solid—animal byproduct	24% calcium. Less effective in soils with pH > 7.
Colloidal phosphate	0–2–2	Very slow	Solid—rock (mined)	Total phosphate is 20%. Less effective in soils with pH > 7.
Cottonseed meal	6–1–2	Medium	Solid—plant byproduct	May contain pesticides. Pesticide-free is also available.
Feather meal	12–0–0	Slow	Solid—animal byproduct	Long-term nitrogen supply.
Fish emulsion	5–2–2	Medium	Liquid—animal byproduct	Can include micronutrients and sulfur.
Fish meal	10–6–2	Medium	Solid—animal byproduct	Micronutrient source.
Fish powder	12–0.25–1	Fast	Liquid—animal byproduct	Micronutrient source.
Greensand	0–1.5–5	Very slow	Solid—rock (mined)	Micronutrient source.
Kelp meal	negligible	Slow	Solid—plant (harvested)	Micronutrient source.
Kelp powder	1–0–4	Fast	Liquid—plant (harvested)	Micronutrient source.
Liquid kelp	negligible	Fast	liquid—plant (harvested)	Micronutrient source.
Potassium magnesium sulfate	0–0–22		solid—rock (mined)	Magnesium (11%) sulfur (22%).
Rock phosphate	0–3–0	Very slow	Solid—rock (mined)	Total phosphate is 32%. Less effective in soils with pH > 7.
Soybean meal	7-2-1	Medium	Solid—plant byproduct	Builds organic matter.

* The specific guaranteed NPK analysis can vary with fertilizer, so check the fertilizer bag label to get the up-to-date values.
** Release time: Fast = immediate, 1 month; medium = 1–4 months; slow = more than 4 months.
Specific fertilizer data compiled from several sources.

LIME AND FERTILIZERS FOR SECONDARY NUTRIENTS

Lime- and sulfur-containing materials are valuable fertilizer sources for the secondary nutrients, calcium, magnesium, and sulfur.

Fertilizers	Estimated Nutrients	Notes and Uses
Limestone (calcium carbonate)	25–40% calcium	Raise pH of acid soils.
Dolomitic limestone (calcium magnesium carbonate)	19–22% calcium, 6–13% magnesium	Raise pH of acid soils. Mg fertilizer.
Gypsum (calcium sulfate)	23% calcium, 18% sulfur	Softens clay soils. S fertilizer.
Elemental sulfur	99% sulfur	Lower pH of alkaline soils. S fertilizer. Natural and synthetic sources.
Epsom salts (magnesium sulfate)	10% magnesium, 13% sulfur	Mg fertilizer. Caution due to salt toxicity use as foliar spray. Natural and synthetic sources.

ESTIMATED NUTRIENT VALUES OF CONCENTRATED AMENDMENTS

The following organic amendments, though not legally fertilizers, are concentrated nutrient sources that can be used like a fertilizer. Manures are also highly concentrated amendments and can supply soil nutrients. Because manure nutrients are highly variable, they are not included in the tables below.

Fertilizers	Estimated Nutrients	Notes and Uses
Eggshells	1–2% nitrogen, calcium	Add ground shells to compost or soils.
Coffee grounds	2% nitrogen	Add to compost or sidedress to provide nitrogen.
Wood ash	6% potassium, calcium, magnesium	Raises pH. Supplies K. Add to compost or mix well in soil. Use cautiously to avoid toxic salt buildup. Maximum application of 1.5 lb. per 100 sq. ft.
Oyster shells	33% calcium	Add ground shells to compost or soils.
Urine	1% nitrogen	Dilute to 1:5 urine to water for compost and perennials. Dilute 1:15 for vegetables. Use cautiously to avoid salts or ammonia burn. Don't apply to leaves.

Match the timing of fertilizer release with when plants need nutrients the most. Maximum plant nitrogen demand is during the leafy green stage of plant growth. This period varies from plant to plant. For corn, it is around four to six weeks after planting.

continued from page 122

Choose the fertilizer with a release time that best meets your soil-feeding schedule. Plan ahead to apply fertilizers at the correct time and synchronize nutrient release with plant demand. Nitrogen demand coincides with rapid plant growth during the leafy green stage before flowering, while phosphorus is needed as a kick-start for new seedlings.

Cost

Cost is another important consideration in choosing a fertilizer. When considering cost, remember to use fertilizers wisely, conserving them to avoid overfertilization and waste. Because fertilizers vary in nutrient concentration, we can't directly compare the cost per

TOTAL VERSUS AVAILABLE: READING MORE FROM YOUR FERTILIZER LABEL

The NPK percentages on organic fertilizer labels have specific meanings in regard to release time. The percent of nitrogen indicates total nitrogen in the bag, which includes all the nitrogen that will be released over the course of months to years. Not all of this nitrogen will be available the first year.

The phosphorus and potassium percentages, however, represent only the amounts that are available in the first season. Some fertilizers contain much higher amounts of phosphorus and potassium than labeled, which are released over three to five years. For instance, rock phosphate has an NPK rating of 0–3–0. This means that 3 percent of the crushed rock will be released as phosphorus the first season. Rock phosphate, however, actually contains 32 percent phosphate, which is released over five years. This means that for 100 pounds of rock phosphate, 3 pounds Phosphate (P) are available the first year, with an additional 29 pounds made slowly available over five to ten years.

COUNTING YOUR COSTS: THE REAL PRICE OF FERTILIZER NUTRIENTS

Because nutrient contents vary from product to product, the price of bulk fertilizers can be deceiving. Get the most bang for your fertilizer buck by comparing the true costs of fertilizer per pound nutrient instead of per pound fertilizer. A 50-pound bag of cottonseed meal is $20 cheaper than bloodmeal, but the cost per pound of nitrogen is 24 cents more expensive.

Product	Bloodmeal	Cottonseed Meal
Price (50-lb. bag)	$50	$30
Price per pound fertilizer	$50/50-lb. bag = $1.00/lb. fertilizer	$30/50-lb. bag = $0.60/lb. fertilizer
NPK	12–0–0 12% nitrogen	7–2–1 7% nitrogen
Price per pound nitrogen	$1.00/lb. fertilizer ÷ 12% N= $8.33/lb. nitrogen	$0.60/lb. fertilizer ÷ 7% N= $8.57/lb. nitrogen

pound of bagged fertilizer between products. Instead, we have to convert cost per pound of fertilizer to cost per pound of nutrient (see box). Organic fertilizers often have a higher cost per pound of nutrient applied than non-organic fertilizers. When comparing the costs between organic and non-organic fertilizers, remember to consider added fertilizer benefits. Although more expensive, organic fertilizers have long-lasting, soil-building effects that may be worth the extra cost.

MATCH IT UP

Matching plant needs with fertilizers and amendments is where the rubber hits the road for whole-soil fertility. We want to rely on the living soil and organic amendments to meet garden nutrient needs, but if we've overlooked or underfed an essential nutrient, then our plants and soils will suffer. This is where a little estimation goes along way.

Armed with soil-test reports, plant nutrient needs, and some general information about organic amendments, you can use this section to walk through the process of figuring out how to close the gap between what nutrients our soil has and what nutrients it needs. Follow these step-by-step instructions to fill out the worksheet at the end of this chapter. After accounting for what you need and what you have, this whole-soil fertility worksheet provides a rough estimate of whether and what quantity of fertilizers is needed to fill in the gaps.

ASHES IN THE GARDEN

Some home-produced organic materials, while not legally fertilizers, are concentrated nutrient sources that are used like fertilizers. They do little to build organic matter, but they can add missing nutrients to garden soils. Wood ashes are an example of this type of concentrated nutrient source.

Wood ashes, spread on garden soils, add potassium the old-fashioned way. Spread ashes in an even, thin layer to avoid problems with salt build-up.

You may have heard tales of your grandparents collecting wood ash to put on the garden. Used this way, wood ash is good source of potassium, calcium, and magnesium. By adding these nutrients, it can also reduce soil acidity. Nutrient concentrations vary depending on temperatures and the type of wood burnt. If you have a wood stove, collecting ashes to broadcast around the garden puts them to good use. Lawns particularly benefit from a sprinkling of wood ash and are less susceptible to adverse effects from over-application.

Wood ash should not be used if soils tend to be alkaline or high in salts. Also, check the source of your ash. Ash from coal, pressure-treated or painted lumber, and garbage are not safe to use in the garden. Because of its high salt concentration, be careful how you apply and store wood ash. It leaches quickly, so keep it covered and dry until ready to use.

For low-potassium soils, apply wood ash at a rate of 1 to 1.5 pounds per 100 square feet. When using, make sure you spread an even, thin layer on the surface of the soil or the top of the lawn. If piled too thickly, wood ash concentrates salts and impedes plant growth. To avoid nutrient loss, do not apply nitrogen fertilizer directly after wood ash. Wood ash also may inhibit seed germination, so give it some time to mellow before planting.

Because whole-soil fertility relies on the living soil, this method is only applicable for gardens that are actively building organic matter. Also be aware that this method is based educated guesses about what nutrients the living soil provides.

What the Soil Needs:

Our first step is to estimate plant and soil needs. For this, we use the following:

1. **Plant Nitrogen Requirements**

    Soil nitrogen is not confidently measured by the nutrient soil test. For this reason, we don't use the soil test to determine nitrogen needs. Instead, we base nitrogen needs on what a plant will actively use during a growing season. See the table above

YEARLY PLANT NITROGEN NEEDS		
Garden Type	Plants	Annual Nitrogen Need (pounds per 1,000 sq. ft.)
Landscape	Native plants or dry gardens	0–1
	Established ornamentals	1–2
	Flower beds and new gardens	2–4
Lawns*	Fast-growing lawns (high maintenance)	3–4
	Slow-growing lawns (low maintenance)	1–2
Berries	Strawberries and cane berries	2–3
Fruit Trees**	1 year old	0–1
	2 years old	2.5
	3–5 years old	2.5–3.25
	6–7 years old	3.25–5
Vegetables	Low: beans, peas	1-2
	Medium: lettuce, tomato, carrot, beet, melon, squash, potato, celery, pepper, spinach	2-3
	High: onion, leek, garlic, sweet corn, asparagus, broccoli	4-6

*Reduce nitrogen needs for lawns containing clovers. When clippings are returned to lawns, use low end of range. When irrigated, use high end of range.

**Average values for apples, pears, and plums from Oregon State University extension publication EC1503. Fruit trees require different fertilization depending on variety, rate of growth, and age. Consult orchard guides for more specific information.

Don't forget the Gospel of Nitrogen. If using these recommendations without a whole-soil fertility plan, apply nitrogen in several split applications. Generally, don't apply more than 1.5 pounds per 1,000 square feet of quick-release nitrogen in a single application.

to estimate annual plant nitrogen needs in pounds per 1,000 square feet of garden. Consult gardening guides for more nitrogen needs of plants not listed below. Enter this value into Column A of the Whole-Soil Fertility Worksheet on page 135.

2. **Soil-Test Report**

   Use the soil-test report to identify other nutrient deficiencies. We want to pay particular attention to phosphorus and potassium.

   • Phosphorus: If soil phosphorus levels rate as "very low" to "medium," enter the P fertilizer recommendation from your soil-test report in Column B of the worksheet. Enter this value as pounds per 1,000 square feet (see the box at the end of this chapter for quick conversions). If your test report makes recommendations based on total phosphorus (P), rather than phosphate (P_2O_5), multiply this number by 2.3.

- Potassium: If soil potassium levels rate as "very low" to "medium," enter the K fertilizer recommendation from your soil test report in Column C. Enter this value as pounds per 1,000 square feet (see the box at the end of this chapter for quick conversions). If your test report makes recommendations based on total potassium (K), rather than potash (K_2O), multiply this number by 1.2.
- Others: Note any other nutrients or micronutrients that rate as "low" or "very low" on the soil-test report. Enter the fertilizer recommendation for any other nutrients in Column D.

What the Soil Has

The organic portion of the soil is a storehouse of nutrients not measured in the soil-test report. The more years that organic amendments have been added to the soil, the greater the organic soil's nutrient-supplying power.

Estimating nutrients available from the organic soil and amendments requires complex measurements and calculations. Recently, a number of resources have been developed to help organic farmers and others who need a high level of accuracy do just that. These are listed in the Additional Resources section at the end of this book. For most of us, however, rough estimates are enough to get started. Regular soil tests and our own observations will give us the feedback we need to adjust our estimates up or down. Use the estimates below to fill out the remainder of the Whole-Soil Fertility Worksheet.

1. **Soil Organic Matter**

   Soil organic matter is a long-term source of nutrients. Every year, a fraction of it decomposes to release NPK, secondary nutrients, and micronutrients. As organic matter in the soil accumulates, the soil's nutrient-supplying power increases. If organic amendments have been added to the soil for the last three or more consecutive years, use the soil test report to estimate the nitrogen it supplies.

   Percent Organic Matter from the Soil Test Report
   A soil-test report will measure the percentage of organic matter of your soil. Estimate the nitrogen released by soil organic matter every year with the following calculations.
   - Divide the percentage of soil organic matter on your soil test report by 100
     — Example: 2% organic matter ÷ 100 = 0.02
   - Multiply the number above by 67.5
     — Example: 0.02 × 67.5 = 1.35 lb. N

- Enter this number in Column A of the Whole-Soil Fertility Worksheet. This is the number of pounds of nitrogen from organice matter per 1,000 square feet of garden that plants can use over a growing season.

2. Cover Crops

Legume cover crops release nitrogen quickly, so incorporating legume cover crops can reduce fertilizer nitrogen requirements. Nitrogen supplied by legumes can vary widely. A productive crop can provide 1 to 2 pounds nitrogen per 1,000 square feet for the coming season. This assumes that the legume cover crop was lush and thriving. It also assumes that the crop was killed and incorporated at the optimum time (see Chapter 4). Use the following estimates on the Whole-Soil Fertility Worksheet:

- Enter a value of 1.5 lb. N in Column A for a dense legume cover crop. Enter a lesser value for a less dense crop or a crop killed before or after the optimum time.

A dense, green stand of red clover has optimum nitrogen just as it is starting to bud. If killed and incorporated at this stage, add a nitrogen value of 1.5 pounds per 1,000 square feet or higher to the soil fertility worksheet.

- If you've used a cereal-legume mixture for a cover crop, enter .75 lb. N in Column A (half of what you would enter for a pure legume stand).

  For more refined estimates, which require weighing the cover crop, see the detailed resources at the end of this chapter.

3. Fresh Manures

Fresh manures are a highly concentrated nutrient source. One of the drawbacks of manures is the difficulty of estimating their nutrient content and availability. As with other organic amendments, lab tests in conjunction with resources listed at the end of this book can help pinpoint exact numbers. Add manures based on availability and recommendations in Chapter 4.

For the purpose of whole-soil-fertility planning, we need rough estimates for the nutrients added by manure applications. In addition to nitrogen, manures are a concentrated source of phosphorus and potassium. They are so concentrated, in fact, that if we apply enough manure to meet plant nitrogen needs, we often apply two to three times the amount of potassium and phosphorus that we need. Over time, this can cause problems due to excessive fertility and salts.

The table on the opposite page provides a starting point for estimating manure nutrients. These are compiled from average published values and assumptions. These estimates assume that the material is fresh, does not have bedding, was handled correctly, and was incorporated within twelve hours of spreading. If this is not true, then reduce the amount of nitrogen entered on the worksheet. Be aware that your particular batch of manure may differ significantly from the numbers below. Use these estimates to fill out the Whole-Soil Fertility Worksheet on page 135:
- Find the type of manure used and multiply each number (N, P_2O_5, and K_2O) by the number of inches actually spread over 1,000 square feet.
- Example:
  ¼ inch of poultry manure was spread over 1,000 square feet of garden.
  N = ¼ × 15 = 3.75 lb.
  P_2O_5 = ¼ × 35 = 8.75 lb.
  K_2O = ¼ × 36 = 9 lb.
- Enter these numbers in Columns A, B, and C of the Whole-Soil Fertility Worksheet.

4. Composts and Other Amendments

Composts and other organic amendments, such as crop residues or fresh yard debris, build organic matter in the soil. This is important, as building soils

ESTIMATED MANURE NUTRIENT CONTENTS

Manure nutrient contents vary widely by animal and from batch to batch. The following estimates, based on average published values, give a rough guess of how much NPK your manure supplies. Composted manures or manures stored and handled improperly will have considerably lower nutrient values.

	Pounds nutrient per 1-inch spread over 1,000 sq. ft.		
	Nitrogen (N)	Phosphate (P_2O_5)	Potash (K_2O)
Horse (without bedding)	0.5	3.5	12
Sheep	6	5.5	27
Rabbit	3.5	6	11
Beef	3	3.5	13
Poultry	15	35	36
Dairy (dry)	20	25	33
Poultry (composted)	2	2.5	15

*1-inch application over 1,000 square feet takes about 3 yards of material.
**A 1-inch application over a 1,000-square-foot application rate is too high for most manures but is an easy number to use for worksheet calculations.

provides soil nutrients in the long term. In the short term—that is, the growing season after they are applied—the nutrient release from these materials is minimal and does not replace fertilizers. For this reason, these additions are not used for whole-soil fertility calculations. Because they build organic matter over time, their nutrient-supplying contribution is accounted for under No. 1, Soil Organic Matter, page 130. Even though these materials are not used for whole-soil fertility calculations, they are a keystone in building whole-soil fertility, as their effect is cumulative. Adding these materials creates long-lasting, sustainable soil fertility while accruing the holistic benefits of organic matter in a living soil.

WHAT DOES IT ALL MEAN?

The numbers are in! We've used our best guesses to fill out the Whole-Soil Fertility Worksheet. Now what does it all mean? Let's look at the following example to find out.

- I'm planning on a slew of summer veggies: tomatoes, peppers, and squashes. From the nitrogen tables, I see these crops need 2 to 4 pounds N per 1,000 square feet. I split the difference and enter 3 in Column A.
- My soil-test report tells I have medium levels of Phosphate (P) and Potash (K). It suggests applications of 2.5 pounds per 1,000 square feet for each. I enter 2.5 in Columns B and C.

- My soil-test report says I have 3 percent organic matter. If I multiply 0.03×67.5, I get 2 pounds N per 1,000 square feet. I enter this in Column A of my worksheet.
- In the late spring, I tilled in red clover that I planted the previous fall. I was a little late incorporating it. Since it had already started to bloom, I reduce the cover crop estimate to 1 pound N per 1,000 square feet. I enter 1 into Column A.
- I got enough horse manure from my neighbor's stalls to apply ½ inch across my garden. I multiply the values from the manure table by .5 and enter the results in Columns A, B, and C.
- I subtract what I have from what I need. From this, I get −0.25 N, 0.75 Phosphate (P), and −3.5 Potash (K). This means that I have .25 pound more N and 3.5 pounds more K than I need, but I'm still lacking a .75 pounds of phosphorus per 1,000 square feet. I'll add a phosphorus-rich source, such as bonemeal, to meet this season's phosphorus requirements (see box on how to calculate fertilizer amounts).
- Because I'm way over on K fertilization, I'll continue to watch soil test levels of K over the next few years. If they continue to increase, I'll stop using manures and opt for other nitrogen sources. If I continue to build the organic matter in my soil, I might gradually estimate a greater and greater percentage of N supplied by the soil (Column A). I'll keep watching my plants to see how they respond to less nitrogen fertilizers.
- Finally, I take a look at Column D. My soil test report says I'm low on sulfur. Because I've added horse manure and have a fairly high soil organic matter content, I assume that this will meet the plant sulfur requirement over the growing season. I'll carefully observe my plants for signs of sulfur deficiency.

Make It Real: Converting Fertilizers Needs to My Fertilizer Bag

By hook or crook, I've figured out that I need to add ¾ lb. (0.75 lb.) Phosphate (P) for every 1,000 square feet of my garden. How do I do this?

1. **Choose Your Fertilizer**

   I scan the fertilizer table and see that bonemeal provides a concentrated form of phosphorus without adding fertilizers that I don't need. I buy some bonemeal at the garden store. It has a fertilizer-analysis tag of 3-15-0.

WHOLE-SOIL FERTILITY WORKSHEET

		Column A	Column B	Column C	Column D
		Nitrogen	Phosphate, P$_2$O$_5$	Potash, K$_2$O	Other nutrients (notes):
		(lb./1,000 sq. ft.)	(lb./1,000 sq. ft.)	(lb./1,000 sq. ft.)	
		What the soil needs			
Plant and soil needs	(+)	3	2.5	2.5	*Sulfur reported as low*
		What the soil has			
From soil organic matter	(-)	2			
From this year's cover crop	(-)	1			
From this year's manure addition	(-)	0.25	1.75	6	
Total needs		−.25 (over)	0.75 (needed)	−3.5 (over)	

2. **Convert Pounds Nutrient to Pounds Fertilizer.**

    I know I need 0.75 lb. Phosphate (P), but how much bonemeal do I need to add? From the fertilizer tag, I know that my bonemeal is 15 percent Phosphate (P). Here's the conversion:

    $15\% = 15 \div 100 = 0.15$

    $0.75 \div 0.15 = 5$ lb. bonemeal needed per 1,000 square feet of garden

3. **How Big is My Garden?**

    My actual planting space consists of two rows that are 4 feet wide by 80 feet long. $2 \times 4 \times 80 = 640$ square feet. I multiply the bonemeal I need by my actual square footage over 1,000. It looks like this:

    5 lb. bonemeal $\times (640/1,000) = 3.2$ pounds bonemeal

    Hooray! I need to spread 3.2 pounds bonemeal across my planting area. I can do that. This provides me with the extra Phosphate (P) my garden needs. It also adds a little extra nitrogen (3.2 lb. bonemeal $\times .03$ N $= .1$ lb. N, to be exact).

FERTILIZER APPLICATION

Congratulations! You now know what fertilizers you need to apply in order to meet soil-nutrient needs. The next step is applying it. Getting the most out of fertilizers depends on not only what you use but how you use it. This includes the method and timing of fertilizer application.

A push-behind rotary spreader uniformly broadcasts fertilizers over large areas, a time and money saver when preparing new gardens or restoring lawns.

How to Apply

Fertilizers can be broadcast evenly across the soil and tilled into beds before planting. This is a simple and all-inclusive way to ready vegetable gardens for the next season or to prepare the soil before planting a new lawn or landscape garden. Banding fertilizers in trenches below new seedlings provides a concentrated dose of nutrients in the plant root zone, where the seedlings need it most. This is particularly useful for slow-to-move phosphorus, a nutrient that is particularly critical in seedling establishment. For already established perennials or for annual gardens that need an in-season boost, sidedressing fertilizers is an effective way to provide nutrients to individual plants. Certain liquid fertilizers, such as liquid fish emulsion and compost tea, are designed for foliar application. This provides an immediate nutrient boost to plants during the growing season to correct for overlooked deficiencies.

Broadcast Fertilization

To broadcast fertilizers, spread fertilizers over the soil surface at the calculated amount. To incorporate, lightly till fertilizers into the soil 2 to 3 inches deep or water the nutrients into the soil with a soaking irrigation. Take care to incorporate nitrogen fertilizers soon after spreading to minimize nitrogen losses.

For small areas, spreading can be done by hand, with a bucket or other small container. For larger areas, various spreaders are available from garden stores, such as drop or rotary spreaders. Rotary spreaders can cover large areas fast, but the evenness of application varies. Drop spreaders usually distribute fertilizers with greater uniformity, although there is some danger of overlapping fertilizer bands. When this happens, overapplication of fertilizers in specific spots can cause salt problems that impede seed germination or growth.

Broadcasting by hand, though less uniform, is an easy way to spread fertilizers over smaller areas. Avoid over- or underfeeding by calibrating a bucket to know how much fertilizer to apply.

Regardless of spreading technique, you'll need to calibrate the spreader. For store-bought spreaders, refer to the instructions and the fertilizer bag to get the right amount of fertilizer per square foot. If spreading by hand, calibrate the container to know how much fertilizer it holds. This can be done with a kitchen scale or by eyeballing the fertilizer bag and estimating how much volume your container holds.

Banding

Banding fertilizers puts nutrients right where the plants need them—at the root zone. New seedlings need a good supply of phosphorus to get established. If your soil is low in phosphorus, banding your phosphorus fertilizer may help new plants get a head start. Organic phosphorus sources, however, work better if broadcast throughout the soil. Potassium can also be applied with banding. The danger of banding is that concentrated fertilizer salts can actually negatively affect plant growth and germination. Because of this, it's important to lay the band close, but not too close, to new seedlings.

Use a trowel to dig a trench around 3 inches deep. Lay your fertilizer in the trench and cover. Plant new seeds at the correct depth, in rows above and to the side of fertilizers. Make sure there are about 2 inches or more between the fertilizer and the new seeds.

Sidedressing

Sidedressing is used to apply fertilizers on established perennials or to give an in-season application to annual gardens. This is particularly useful for sandier soils, where the season's nitrogen is best applied in two split applications (see below sections). In addition to fertilizers, sidedressing is a tried-and-true way to apply compost, manures, and nutrient-dense amendments, such as coffee grounds or eggshells.

To sidedress perennials or annuals, apply the fertilizer or amendment close to growing plants. To keep from harming plants, avoid direct contact with stems, leaves, or roots. Scratch the fertilizer lightly into the soil surface, without disturbing roots. Apply irrigation to allow nutrients to absorb into the soil.

Foliar Feeding

Foliar feeding applies fertilizers directly to plant leaves. These are immediately absorbed for a rapid plant growth response. This can correct nutrient deficiencies, particularly nitrogen and micronutrient deficiencies, during the growing season. Follow the directions on liquid fertilizer mixes and make sure that fertilizers are diluted enough not to burn or damage leaves.

Sidedressing with a nitrogen source, such as coffee grounds, manure, or other nitrogen fertilizers, can provide a mid-season nitrogen boost.

Foliar feeding with dilute liquid fertilizers gives plants a quick boost during the growing season. This helps with nitrogen or micronutrient deficiencies that weren't corrected before planting.

The need for in-season nitrogen can result from insufficient initial applications, cold soils causing slow nutrient release, or late-season nutrient needs of long-season plants. If annuals are slow growing and stunted, with yellowing leaves, a foliar nitrogen feeding is a quick Band-Aid solution.

Many gardeners use foliar feeding to apply compost or manure tea. Specific recipes and processes for making these teas vary, but they are essentially created by soaking compost or manures in water and using the leachate as an amendment. It is unclear whether compost or manure teas supply significant nutrients for plants. Other benefits attributed to teas include reducing plant pathogens and inoculating soils with beneficial microorganisms, though the scientific evidence for these claims is mixed. Just as the nutrient contents of composts and manures vary from batch to batch, the nutrient content of teas certainly varies as well. To foliar feed with compost or manure tea, check the spray on a few plants to make sure there are no adverse effects before applying it across your garden.

Another reason for poor growth could be a lack of micronutrients. Apply a micronutrient-rich foliar fertilizer, such as liquid seaweed, to see if the plants respond. A response to micronutrient foliar feeding could indicate deficiencies or pH problems in the soil. To fully evaluate this, make sure to get a soil test and address any underlying problems. Foliar fertilizers are a quick, though costly, solution. They will not, however, address the underlying reasons for deficient micronutrients.

When to Apply Fertilizers

When to apply fertilizer is also important for maximizing how well plants use the added nutrients. There are a few things to consider about getting the timing right. These include soil type, fertilizer release rate, and what you are fertilizing. This can be simplified in a few simple rules.

Rule 1: Remember the Gospel of Nitrogen

If soils are sandy, apply nitrogen twice. Chapter 1 discussed how to determine soil type and textural class using the USDA textural triangle or a soil test report. Soils with sand in their textural class name (i.e., sandy loam, sandy clay, or sand) are, obviously, considered on the sandy end of the soil spectrum. These soils don't hold onto nitrogen well. Because of this, it's better to add the season's nitrogen requirements in two, or even three, split applications. To do this, divide the nitrogen fertilizer needed by the number of applications you plan. Apply at the beginning and middle of the season. This ensures that the nitrogen added gets used by plants and doesn't drain away to pollute the water table. When using chemical fertilizers, split applications are recommended regardless of the soil type. Long-season, heavy nitrogen feeders, such as corn, also benefit from receiving their nitrogen ration in two or three applications throughout the season.

Rule 2: Count Backward

The fertilizer tables in this chapter, in addition to specifications on the fertilizer bags themselves, give information about how quickly fertilizer nutrients are released. Use this information to count backward from the garden's highest fertilizer demand. Some phosphorus is needed right away, while nitrogen needs peak during the rapid growth period that starts two to four weeks after planting.

For preparing new beds, use these general rules of thumb. Apply slow-release fertilizers in the fall before planting and moderate release fertilizers two to four weeks before planting in the spring. Apply quick-release, chemical fertilizers right before planting. Better yet, use these in split applications throughout the season to supply fertilizers as needed.

Rule 3: Add Lime in Advance

Lime takes a while to break down in the soil and even longer to undergo the chemical reactions that balance pH. Because pH is critical in making nutrients available to plants, make sure to prepare new beds by applying lime well in advance of next season's planting. The time it takes for lime to react depends on the size of the material. Fine, powdered materials react more quickly in the soil. Coarser, granule- to gravel-sized materials are cheaper and have a pH balancing effect over several years. Lime is best applied and lightly tilled into the top 6 inches of soil the fall before planting. If incorporating amendments, cover crops, or other organic residues, spread the lime evenly over the material and incorporate it in one go. At a minimum, apply lime four weeks before any new planting.

Fertilize fruit trees with a sidedressing of manure in the late winter, before leaves start to bud. Trees make early use of nutrients to grow roots, shoots, and more productive fruits in the coming season.

Rule 4: Fertilize Perennials Early

A flush of nutrients to trees, shrubs, berries, and other perennials causes them to put on new growth. While this is important for productivity early in the season, it can interfere with dormancy if applied too late. This results in winterkill of fresh and tender perennial shoots that can damage the plants. It also prevents them from conserving nutrients to get through the winter. The exact date varies from region to region, but do not apply nutrient-rich materials such as fertilizers or manures after late spring. Trees, shrubs, and other perennials get the most benefit from a late winter or early spring fertilizer application.

Rule 5: Fertilize Lawns in the Fall

Lawns with cool-season grasses do best when fertilized with phosphorus and potassium in the late summer or early fall. This allows them to produce new roots before the winter freeze, which will give them a burst of growth in the spring. Lawns with warm-season grasses have slightly different requirements. Fertilize these in split applications throughout the spring and summer growing seasons.

CHAPTER 6

THE DRESSING ON TOP

MULCH, WATER, AND WEEDS

After the important work of growing, amending, and fertilizing a living soil, this chapter looks at the dressing on top: how and when to add organic mulches and how and when to manage garden water. These two activities are responsible for making sure garden plants drink their fill, while minimizing competition from weeds and pests. Moreover, good mulch and water management create the juicy soil habitat a living soil needs to thrive. By following the guiding principles of keep the soil covered and the Goldilocks rule of not adding too much or too little water, we create the optimum conditions for air, water, and shelter in the soil environment. Paying attention to these basic garden tasks dramatically multiples the beneficial effects of building soils with organic matter.

COVER THE SOIL

Walk out into a meadow or forest and look at the ground. What you won't see is the soil. In nature, the ground is covered. Living plants, recently fallen leaves, or a thatch of dead grasses form a soft cap on the hard earth.

This cover serves many purposes. It protects the soil from erosion, runoff, and rainfall impact. It regulates temperature and intercepts the drying rays of the sun. It conserves water, while at the same time allowing more rainwater to be caught and held by the soil. It provides nutrients, which slowly leach into the soil over time.

Mulch provides shelter and water in the soil environment. Whatever form it comes in—whether living, newly dead, or long-dead—organic matter covers the soil to give soil microbes a good home.

All of these effects serve one critical function: creating a good home for soil organisms. This natural cover goes a long way to meeting the living soil's requirements for shelter, water, air, and food. Without cover, the top ½ inch of soil becomes lifeless when too dry, wet, or cold. Over large areas (like our gardens), this amounts to the loss of quite a bit of living soil activity.

We can copy this natural process by covering garden soils with mulch. This simple cover has one of the most dramatic effects on garden productivity. In my own garden, mulching is one of the most critical tasks I perform. Even when stretched thin for time or resources, I rarely skip the mulch. For the investment of time required to spread (or grow) a mulch, the paybacks are enormous.

A woodland is self mulching. Nutrients are recycled through fallen leaves. Under this shelter, living fungal mats cover the soil surface, while bacteria and other soil animals take advantage of the moist habitat.

An uncovered soil is vulnerable to weather events. A rainstorm can carry away precious topsoil by the force of water. Mulch helps slow and capture water at the soil surface to prevent soil loss.

Mulches are intimately connected to the most time-consuming garden maintenance tasks: watering and weeding. By increasing the amount of water held in the soil, mulch reduces the frequency with which I need to irrigate. The water I do apply is used more efficiently, as mulch helps plants develop more extensive root systems. Mulches keep the soil surface soft and moist and prevent the formation of hard-to-manage surface crusts. A thick mulch smothers unwanted weeds and exponentially reduces the time I spend removing them. Most weeds that do come up through thick mulch are shallowly rooted and easy to pull.

Most importantly, mulches build organic matter in the soil. This happens directly over time as burrowing organisms bury organic mulches and add food to deeper soil layers. Mulch also has a more immediate, though indirect, effect on organic matter by building good homes for the living soil. In this hospitable habitat, soil organisms create the soil conditions that hold more and more organic matter.

Mulches have their limitations as well. They can keep the soil too damp and moist, an environment that can harbor slugs and other pests. Avoid thick mulches on waterlogged soils that already have drainage problems. For early spring planting in clay soils, remove mulches to allow the soil to warm. Mulched perennials develop strong root systems at the soil surface, which are susceptible to damage if the mulch is removed. Keep perennial gardens covered to protect these strong but sometimes shallow roots.

Types of Mulch

I like to mulch with a range of different materials. This is one of the easiest ways to get a variety of organic foods into the living soil, while creating a variety of soil habitats. Increasing aboveground diversity increases belowground diversity, making the living soil stronger and more resilient. This satisfies the guiding principle: make your life simple by making your garden complex.

Many of the organic amendments described in Chapter 4 can also be used as mulches. When composts and manures are applied to the soil surface, however, they provide fewer nutrients than when incorporated as an amendment. Some of the best mulches, in fact, are nutrient-poor materials because they stay on the soil surface for a longer time. Low-quality mulches, such as wood chips, bark, sawdust, and wheat straw, decompose slowly to provide long-lasting benefits.

When choosing a mulch, availability is the most important factor. Use what's readily at hand to cover and shelter the soil. Because I'm always looking for ways to add organic

I'll let my straw bales sit for a year to rot into the perfect vegetable garden mulch. When partially decomposed, straw is easier to spread and doesn't tie up nutrients as it rots into the soil.

food to the soil, I prefer to mulch with organic materials. Mulch size, nutrient power, and water absorbency are other characteristics to consider when choosing a mulch material. Different materials also have advantages and disadvantages, depending on your reason for mulching.

If mulching for water conservation, choose a material that is loose and coarse textured. This provides a surface that catches water easily, while allowing water to percolate easily to the soil below. A material that absorbs too much water is not a good choice. We ultimately want the soil, not the mulch, to absorb water. In the perennial garden, I like to use chipped arborist mulch, which conserves water while adding some slowly decomposing organic material. In the vegetable garden, my favorite mulch is rotting straw bales, which are easy to spread and use.

Finely chipped or shredded material, such as bark dust or sawdust, is less preferable. These small particles can crust at the surface and prevent water infiltration. Similarly,

If you can chip it, then you can mulch it. Use a chipper-shredder to turn landscape trimmings into valuable garden mulch that adds organic matter, boosts garden productivity, and saves you valuable time and effort.

FIND A MULCH TO FIT YOUR NEEDS

Mulches come in all shapes and sizes to fit different gardens and garden purposes. You may choose a mulch material for its ability to conserve water, control weeds, add organic matter, or create an attractive landscape.

Material	Description
Straw	Straw is harvested in bales before seedheads emerge, provides good water infiltration, and is easy to spread and remove. Aged, rotting straw supplies more nutrients than fresh straw. Most straw contains some weeds seeds. Straw weeds are easily pulled if caught early. Lifespan: one season.
Hay	Hay is harvested in bales that contain weedy seedheads. To reduce weed seed problems, use hay harvested before seed maturity. Legume hays, such as alfalfa, are excellent nitrogen sources. Hay is more difficult to spread and more expensive than straw. Lifespan: one season.
Bark mulch	Bark mulch is commercially available from landscape suppliers. It's a long-lasting material that is excellent for weed control and water conservation. Many gardeners find it gives a neat, clean appearance in the perennial garden. Lifespan: two to four years.
Wood chips	Chipped waste from tree trunks is good for water conservation and weed control and great for mulching pathways, perennial beds, or fruit trees. Lifespan: one to two years.
Arborist mulch	Chipped and shredded brush and tree trimmings are available from local arborists; you can also rent or buy a chipper-shredder to make your own. This is a great mulch for weeds and water. It provides some nutrients, depending on the type and greenness of the starting material, but it's less uniform and more difficult to apply than chips or bark. Lifespan: one to two years.
Leaves	You can collect leaves for free in your neighborhood. They are a great source of nutrients and organic matter. They conserve water and prevent weeds. Shred leaves to prevent the formation of water-repellent mats. Check periodically to ensure that water infiltrates through the mulch. Lifespan: one season.
Grass clippings	Leave grass clippings in place on the lawn for a natural mulch. For water and weed control in the garden, let clippings dry before use. Thick layers of green clippings can repel water. Use a thin green layer for a nitrogen boost under carbon-rich mulch. Use thick applications for nitrogen in sheet mulches. Lifespan: one season.
Seed hulls	A variety of seed hulls from agricultural processing are available commercially. Examples include hazelnut hulls and cocoa bean hulls. Prices, uses, and longevity of these materials vary. When plentiful and affordable, seed hulls are a good perennial mulch material.
Pine needles	Pine needles provide good weed suppression and moisture control. They are used in commercial landscaping because of their clean, neat appearance. Interlocking needles are good for erosion control. Lifespan: two to three years.
Cardboard	Cardboard is a highly effective weed-smothering mat. Use for pathways or the first layer of a sheet mulch. Be sure to wet thoroughly and cover with a heavier material to prevent it from blowing away. Lifespan: one year if moist.
Compost	Compost conserves water while adding some nutrients and organic matter to the soil. It's least effective for weed control. Its black color can help soils heat up in spring. Lifespan: one year.
Sawdust	Sawdust has similar properties to wood chips, but smaller-sized particles decompose more quickly. It may crust to impede water infiltration, but it's great for weed control. Use it for pathways or shrubby perennials. It's not appropriate for annual gardens. Lifespan: one year.

For a smothering mulch, lay cardboard in pathways or sheet-mulched garden beds. Wet the cardboard and then cover with wood chips or gravel.

dense and nutrient-rich materials, such as grass clippings or leaves, may form water-repellent mats. If using these materials, check periodically to make sure water is passing through easily. If not, loosen the mulch surface with a rake. To prevent mats from forming, allow grass clippings to dry thoroughly and shred leaves before use.

When mulching for weed control, on the other hand, we actually want to block water (and sunlight) to prevent seed germination and seedling growth. For this purpose, a thick mulch of finer material, such as sawdust, actually works the best. To eliminate weeds in an area, completely block water and sun with a smothering layer of cardboard or newspaper. Obviously, when mulching for weeds, don't use weed-containing materials, such as hay, manures, or unfinished compost.

Several nonorganic materials also work as landscape mulches. Because they don't provide the benefit of adding organic matter to the soil, I generally avoid them. These nonorganic options include plastic mulches, landscape fabrics, and gravels. Personally, I don't like using plastic or fabrics in my garden. Black plastic degrades in sunlight, turning into a garden mess. Over time, weeds grow up through both types of materials and are more difficult to pull and manage. Sometimes, I do use gravels for pathways and dry landscape gardens. Weeds will eventually come up through gaps between the gravel particles; but in low-water situations, they can be easily managed with a flame weeder.

Straw mulch in vegetable gardens can serve many purposes. Pile heaps of straw on top of potato vines as they grow to build the garden bed vertically and increase potato yield. Tubers that form in the straw are easy to harvest by hand.

Using Mulch

When I spread mulch in the garden, I'm trying to accomplish many things at once. Conserving water, suppressing weeds, and feeding the living soil are at the top of the list. Insulating soils and plants from freeze-thaw, preventing erosion on pathways or sloping gardens, and protecting tender vegetables and seedlings are additional concerns that can be addressed with mulching.

Generally, mulches are applied in a thick, 2- to 4-inch layer. Thinner mulches will shift to create bare spots, while deeper mulches may prevent water from reaching the soil. For weed control, aim for a minimum of 3 inches. Thicker mulches are needed for erosion control on hillsides and sloping land.

In cold winter areas, mulches can help with cycles of freeze and thaw that cause the soil to heft and heave. Mulch after the first freeze to keep the ground insulated from temperature fluctuations. You can also pile mulch around frost-sensitive vegetables to protect them from frost and extend the growing season.

In the vegetable garden, mulches protect sensitive vegetables from the sun and help seeds germinate. After sowing, cover beds with straw mulch to keep the soil moist. Large-seeded annuals, like beans or corn, will grow up through the mulch layer. For smaller-seeded vegetables, such as lettuce or carrots, periodically pull back the mulch in the first few days after planting. Once seeds germinate, remove the mulch completely until new seedlings are large enough to thin.

Sheet mulching, as discussed in Chapter 4, is primarily used to amend soils by composting in place. It is also a highly effective weed-control method. If I want to eradicate weeds in problem areas of my garden, I'll use a sheet mulch to smother the area while building new soil.

A Living Mulch

When at all possible, I like to let the garden do my work for me. If I can avoid hauling, forking, spreading, or raking, I will. Because of this, I use living mulches whenever I can to cover bare soil. Living mulches work because they compete with weeds for nutrients, sun, and water. They occupy otherwise bare soil that weeds can exploit. Meanwhile, they build organic matter by adding fresh roots and shoots to the soil.

Living mulches compete with desired garden plants as well as weeds. To avoid unwanted competition when planted together, choose plants of different heights and rooting depths. This maximizes the use of sunlight, water, and nutrients. To further reduce competition, maintain a buffer between living mulches and garden vegetables.

I divide living mulches into three main categories: groundcovers, green manures, and intercrops. In addition to controlling weeds, living mulches add organic matter and nutrients, provide an additional crop, and make an attractive addition to the garden. In this way, increasing overall above- and belowground garden diversity also benefits the soil organisms and beneficial insects in your garden.

Low-growing wooly thyme (in foreground of photo) is one of my favorite groundcovers to use as a living mulch because of its low maintenance and great aroma.

Groundcovers are low-growing and often shade-tolerant plants that form an attractive cover in the perennial garden. Once established, they are relatively maintenance free. I like to choose groundcovers that have low water requirements, as they compete less with other garden plants. Unfortunately, many invasive plants are sold as ornamental groundcovers in garden centers. Be sure to check with your local extension service for invasive ornamentals in your state before choosing a groundcover that may become a problem.

As an essential way of amending a thriving garden ecosystem, Chapter 4 discusses green manures in detail. When using green manures in or near other garden plants, reduce competition for sun by choosing varieties with complementary growth habits. You can also reduce competition by staggering the timing of green manures. Plant quick-growing green manures early to cover bare soil and pull up as vegetables mature. Alternatively, plant green manures late, into maturing vegetables, to provide cover after harvest. When you pull, cut, or mow green manures, use them as a handy and nutritious mulch for nearby plants.

Mutually beneficial onions and lettuce increase overall garden productivity when planted as an intercrop. The broad leaves of lettuce are a living mulch that reduces weed competition. Onions shade heat-sensitive lettuce from the sun.

An intercrop is simply a crop grown in proximity to another crop. Instead of growing garden vegetables in homogenous blocks, we can plant them together to increase garden complexity, diversity, and overall garden productivity. As a living mulch, interspersed intercrops also cover otherwise bare soil. For this purpose, a lettuce and onion intercrop is one of my favorite combinations. The broad leaves of lettuce help control weeds between onion rows. Heat-sensitive and shade-tolerant lettuce also benefits from the partial shade of tall onion stems. Though both lettuce and onions can develop deep root systems, lettuce has more fine surface roots that maximize the use of water and nutrients closer to the soil surface.

As with other soil-growing practices, living mulches allow you to exercise your gardening creativity. Armed with the principle of keeping the soil covered, find ways to fill bare garden spots with living garden plants. Observe the effects and learn from your successes and mistakes. Not only does filling these bare spots reduce weeds and increase organic matter, it can give you a chance to add color and texture to your garden. A pretty flower or a new smell, while acting as a living mulch, builds your garden's diversity in a way that combines form, function, and beauty.

WEEDS

Weeds do have their place in the garden. They, too, are living plants that provide organic matter and living cover. A nuisance weed for one gardener may be a garden helper for another. Dandelions, though persistently pulled from lawns, capture deep soil nutrients. In a similar way, unwanted clovers capture nitrogen to increase fertility. When dealing with weeds, I weigh the cost of removing the plant against the actual harm they are causing. Sometimes, they may even be beneficial.

Nonetheless, it is important to control weeds before they control us. As soil growers, we want to minimize disturbance. This means we need to control weeds early in the game, when simple, less invasive methods will do the trick. Invasive weeds spread like wildfire through a garden and greedily feed on rich organic soil. We must mulch, water, or rip these out of existence before they take over.

A flame weeder saves time in the garden. It works particularly well for controlling weeds in pathways or xeriscapes (dry garden landscapes).

GARDEN ENEMIES: HOW TO DEAL WITH INVASIVE WEEDS

In every region, there are some weeds that require extra special attention. How do we know they are invasive? These are the garden pests that keep coming back again and again after we pull them, aggressively overtaking our gardens. They can come up through the thickest mulches or reemerge every time we till. Generally, these are either weeds that reproduce from the tiniest bit of root or rhizome or weed seeds that can last for years and years in the soil. Over time, improving soil health helps garden plants compete with invasive weeds. In the short-term, however, the garden needs your help to eradicate these pests. To do so, get to know your garden enemies by using weed handbooks and consulting local extension services and garden centers.

Although we generally want to recycle all the plant matter from our gardens back into the garden soil, invasive weeds require special care. If pulling weeds that reproduce from root or rhizome, try to get every bit of root out of the soil. For weeds that spread aggressively by seed, be sure to cut or pull stems before seedheads form. Even immature seeds can mature and germinate in the soil. Do not put any part of these weeds in the compost: stem, leaf, root, rhizome, or seed. Lay them in a single layer on a plastic tarp in full sunlight. Turn periodically over the course of a week or two. Once you are sure that they are completely dried and dead, they can be added to compost (if you dare).

Crabgrass

Bindweed

Vinca (periwinkle)

Public enemies No. 1, 2, and 3: Beware of universally invasive plants like crabgrass, bindweed, and vinca. Know which weeds are particularly noxious in your area and get them before they get you.

As we gain a soil's-eye view, it becomes more and more apparent that all gardening activities are connected. Nowhere is this truer than in dealing with weeds. Unwanted weeds compete for water, sunlight, and nutrients with the garden you want to grow. Controlling weeds, by the same token, relies on reducing this competition. To do this, we utilize how and when we fertilize, mulch, and water. Ultimately, we want to use garden plants themselves to self-regulate and control unwanted weeds. Growing healthy garden plants, in fact, is the number-one way to control garden weeds. This, in turn, comes back to growing a healthy living soil.

We also control weeds with mulch and water. Thick mulches, as discussed above, block out sunlight. Water-impenetrable mulches also block out water. How and where we water is also extremely important in weed management. In most cases, more plant-specific watering reduces weed problems. For instance, applying drip irrigation instead of overhead sprinkling delivers water only to the plants that need it. In some cases, when drought-tolerant weeds are a problem, more water actually gives garden plants the edge they need to compete with weeds.

We also want to avoid weed problems because of how they affect organic matter in the soil. When we pull and remove weeds from the garden, we remove organic matter and nutrients from the soil. When at all possible, recycle weeds by adding them to the compost pile, chopping them into the soil, or mulching them on the soil surface. These techniques should not be used after seedheads form or for particularly invasive species that reproduce from roots or rhizomes. Because of the dangers of invasive weeds, it's extremely important to know your enemies when deciding how to treat and manage weeds.

WATER IN THE GARDEN

Water is the lifeblood of the garden. We have all commiserated with shriveled, dried-out plants in the heat of summer. If water-stressed early in life, plants may never fully recover. At the same time, soils that are too wet also stunt plant growth. Because roots need oxygen to grow, they won't grow into waterlogged soils.

What we don't always think about is that soil organisms also need water. If a soil is too dry, then soil life goes dormant waiting for water. If it's too wet, soil organisms are limited by lack of air. Because of this, both over- and underwatering stresses the soil ecosystem and reduces how well it functions. For this reason, the right water at the right time is critical for growing living soils.

The Right Water for Your Soil Texture

We can think of the soil as one big water tank. Remember that half of the soil is composed of empty pore space. This is the soil water tank. When we irrigate, we want to fill, but not overfill, this tank. To do this we need to know (1) how big the tank is, (2) how fast it drains, and (3) how much water is in the soil tank right now.

Soil texture fills in the first two pieces of information. Clay soils have a lot more pore space than sands. This means they hold much more water. A clay soil is like a 5-gallon tank, while a sandy soil is more like a quart-sized water bottle. If plants drink the same amount of water in both, you'll need to fill up the sandy water bottle about twenty times for every one time you fill up the tank of a clay soil. To complicate things, clays are sticky, greedy particles that don't share their water as freely as sands. This means that of the 5 gallons of water in the tank, a clay soil may share only 2 gallons with garden plants.

The Soil "Water Tank"

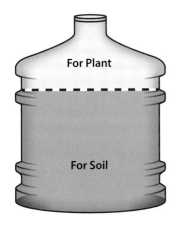

Clay
most water held in the soil

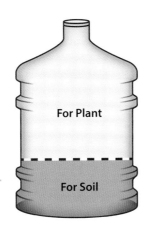

Loam
most water for the plant

Sand
very little water in the soil
or for the plant

Soils are giant water tanks that are filled with rain and irrigation water. The size of the tank and how fast it drains depends on your soil texture. Soils also have a minimum fill level, below which they won't share water with garden plants. Clay soils have larger tanks and store more water, yet they have a high minimum fill level. Sandy soils have small water tanks. They drain very quickly and have low minimum fill requirements. Loamy soils have medium-sized tanks, yet a low minimum fill, meaning they share most of the water they hold with plants.

Loams, with a mixture of particles, hold less water than clays but share it more freely. For this reason, the water tank of a loam soil holds the most water that plants can actually use.

How fast the soil water tank drains is the other important piece of watering information. Sandy soils are filled with large, interconnected highways of pore space, which drain extremely fast. This means that even if you fill the water bottle of pore space, the plants may not have time to drink the whole thing before it slips away. The space between tiny clay particles, on the other hand, is not well connected. Water is caught in these tiny crevices and doesn't drain easily. Because of this, clay-rich soils easily become waterlogged if watered too much or too frequently. This means plants and microbes can't breathe. For these reasons, knowing soil texture is the key to watering the right amount at the right time.

In general, water sandy soils with less water, but more frequently, than clay soils. When we overfill the tank of a sandy soil, the water just drains right out, carrying vital nutrients with it. Fill up the large clay soil water tank with a longer and deeper irrigation, less frequently. To avoid waterlogged conditions, let the tank run low before filling it up again.

Generally, we want to water when the soil tank is between 75 and 50 percent full. How do we know when it's time? We simply ask the soil. Go into the garden, dig down 6 inches to a foot, and take a few good handfuls of soil. If a soil is dry or only slightly moist, then it is time to water. A moist or wet soil can wait longer before the next irrigation. For a sandy soil, this may mean waiting a day or two, while for a clay soil, it could be a week or more. When you dig down to check moisture, make sure that the soil stays moist deep into the rooting zone. Depending on the plant and the season, this can be 2 feet or more. If the soil stays moist to this depth it means you're applying enough water to reach deeper roots.

Use your hands to measure soil moisture. Differently textured soils have a specific feel when they are moist or dry. To truly calibrate your hands and learn more about how to schedule irrigation according to soil texture, see the USDA publication *Estimating Soil Moisture by Feel and Appearance* in the Additional Resources section.

With this sort of hands-on soil moisture check, you can refine your watering schedule—the timing, frequency, depth, and amount of irrigation—to your specific soil needs. Over time, you'll know the schedule that works for you. In the heat of summer, gardens need more water because of evaporation. As plants grow larger and roots go deeper, plants need a deeper, longer watering as well. With mulch, soils need less frequent, yet deeper, watering.

The amount of water needed to refill the soil tank depends on how deep you want water to go. Use the following guidelines as a starting point that you can refine with your own soil estimates. You can determine your soil type using the texture section of Chapter 1.

Feel the soil to know when to water. A dry to slightly moist soil, if it holds a shape at all, will only form a weak ball when pressed together. It also won't leave water stains on your finger. A moist or wet soil, on the other hand, generally can form a stronger ball and leaves water stains on your fingers.

- For coarse-textured soils (sand, loamy sand, and sandy loam) apply less than 1 inch for every foot of depth you want the water to percolate.
- For medium- to fine-textured soils, use your moisture check as a guide. Medium-textured soils (sandy clay loams, loams, and silt loams) can hold up to 2 inches, and fine-textured soils (clay, clay loams, or silty clay loams) up to about 2½ inches per foot of soil depth.

To calibrate your irrigation system, place a plastic container in the watering zone. Mark a line 1 inch from the bottom of the container and time how long it takes to reach this depth. This is the time you will need to run your irrigation to apply the desired amount of water.

Soil texture also affects how quickly water infiltrates the soil. Water passes into sandy soils much quicker than clay-rich soils. You want to apply water slowly enough that it can infiltrate into your particular soil water tank without spilling over. If water pools at the soil surface, then reduce the rate at which you irrigate. This gives the water the time it needs to percolate into the rooting zone.

Water on the Landscape: How It Goes and Where It Flows

Irrigation is only one small part of water management in the garden. Many gardens rely on rainwater alone. Regardless of whether water comes in the form of rain or irrigation, we want to shape our gardens in a way that maximizes water use while minimizing water damage. Watching how water goes onto and where it flows into the garden is the first step in working with water as a part of the whole-garden system. To do this, walk outside in a rainstorm and watch where the water goes and how fast it moves.

In general, we want to catch and keep water by making our gardens into sustainable sponges. To do this, we need to slow the flow. This allows time for water to seep deeper into the soil and reduces its power to erode garden beds. Anything I can do to make the soil surface more complex and rough helps. A coarse mulch or a living cover of garden plants works for this purpose. The more dense and complex my garden, the more water

Swales for a Garden Slope

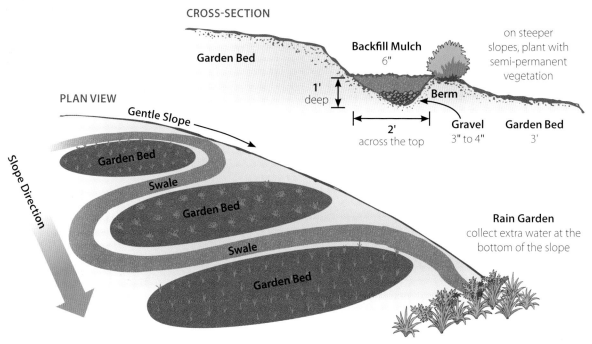

Spread the flow of water in a zigzag pattern across your garden using swales, trenches, or even garden pathways. This technique slows water to refill the soil water tank. On steeper slopes, increase the swales' ability to catch water by filling it with gravel, topping it with mulch, and planting the berm with perennials that hold the soil in place. Use swales to create terraced garden beds and direct water to the plants that need it.

Swales between garden beds spread and direct water, even on gentle slopes. Use this technique to maximize plant water use while minimizing water damage due to erosion or waterlogging.

is retained by the system. A mixed perennial-food garden, for instance, has multiple layers close to the soil surface to slow and capture water.

On sloped ground, swales offer another way to direct and slow water. Swales are a permaculture design technique that spread water across the landscape. If built across the contour of a slope, swales increase the path of water, which slows its speed. Using this technique, water zigzags across a slope to increase the amount of time it is held in a garden and, thus, the amount of water that is used by plants. This technique works even on very gently sloping land. Simply building garden beds perpendicular to a slope, a technique essential for preventing erosion, also acts as a sort of swale by directing water across the contour of a slope. On steeper ground, the bottoms of swales can be backfilled with gravel or covered with mulch and downslope berms planted with either wetland or food forest species, depending on the amount of water you receive. This technique delivers water directly to plants adjacent to the swales while providing more aggressive erosion-control measures.

Water Problems

Many soil problems in the garden come back to problems with water. In my own southern Oregon garden, dry summers, excessively wet winters, and a heavy clay soil are the biggest challenges to soil life. When the soil organisms don't have enough water or air, I can see the effects. Mulch sits dry on the soil surface. Plants don't grow with vigor. Sometimes, drought-loving weeds such as yellow star thistle thrive. In these situations, improving the soil's ability to hold and drain water allows the living soil to grow.

Once you are able to capture water on the landscape and prevent soil erosion on sloping ground, as discussed above, water problems in the garden boil down to two main concerns: excessive drainage (water leaves too quickly) and poor drainage (water stays too long).

Water Leaves Too Quickly

Sandy soils, which drain water like a sieve, are notoriously poor at holding it. This is part of the reason that a dry, sandy beach is much easier to enjoy than a clay muck beach. It does, however, create gardening challenges due to water limitations. Despite this problem, these types of soil also have their advantages. They warm faster in the spring to allow for earlier seeding. They provide ideal conditions for drainage-intolerant plants and crops, such as asparagus, or rooting crops, such as carrots. For sensitive and picky garden plants, sandy soils allow you to control exactly what these plants receive and when they receive it. Because sandy soils drain so fast, water and nutrients are only available as you add them to the soil.

Nevertheless, these soils can create nutrient and water limitations for most crops. To improve these conditions, we want to slow drainage in a sandy soil with the basic soil-building principle we've used throughout the book: adding organic amendments. Because a sandy soil can't do it on its own, added organic matter helps hold on to water, slows drainage, and provides more consistent food and water for soil organisms and plants.

In addition to building organic matter, gardening strategies that capture water are also beneficial for excessively draining sandy soils. These include heavy mulching, building sunken beds, and using swales and contours to direct water. Planting drought-loving species is another strategy for making water-limited soils more productive.

Water Stays Too Long

At the other extreme, poor drainage, leading to waterlogged conditions, poses serious problems for plants and soil life. Anyone who has tried to garden in spring clay muck knows that too much water in the garden is a problem. This messy and unproductive condition can happen for several reasons—often a combination of several reasons. The contours of a garden may lead to a low-lying spot where water collects. Soils may be rich in clays and silts that impede drainage. High water tables and compacted hardpan can also cause waterlogged conditions, regardless of soil texture.

Observation is your primary tool to prevent garden muck. Before you plant, watch how water flows. If you observe water pooling in poorly draining areas, reserve these for accessory garden activities, such as tool storage or compost-making. As an alternative, plant these areas with wetland plants that don't mind wet feet. Using native plants for a natural wetland creates interest and diverse habitat in your garden.

Dig a shallow swale to direct water runoff either away from or toward your planting areas, depending on whether you have too much or not enough water. Line the swale with drainage rock, cover with topsoil, and plant.

A U-bar tiller (also called a broadfork) is a useful hand tool for compacted soils. Use your body weight to force the tines into dense soil layers. Rock the tiller back and forth to open up drainage in compacted soils.

For moderately poor drainage, which is not caused by physical limitations such as hardpan or high water tables, build soil organic matter to enhance soil conditions. Organic amendments, because they are light and porous, fluff up a clay soil to increase pore space and drainage. As soil health increases, soil organisms further improve permeability and reduce compaction as they build homes and burrows.

Sometimes past use and abuse results in compacted hardpans that prevent drainage. This can occur when a soil is tilled repeatedly at the same depth, impacted by heavy equipment, or worked when too wet. It can also result from natural conditions. To deal with hardpan, or extremely compacted soil layers, a deep ripping can sometimes help.

Plant a "rain garden" of wetland plants in a low area that naturally collects rainwater. This takes advantage of the natural flow of resources in the garden ecosystem.

Yes, you read that right; *ripping* is a technical term for pulling a big hook, known as a ripper or a subsoiler, through the soil. Forget what we've learned about building structure and increasing permeability through the gentle habits of organisms. When we rip a soil, we're basically trying to make a deep gash in the compacted layer to let water drain. Deep ripping is a bandage for the underlying soil problem. If soil can drain, then soil organisms can become active and plants can grow. This starts the soil system, which can then continue to improve soil naturally. By the same principle, a less invasive approach is to use a body-powered U-bar tiller to open up compacted layers.

If excess water results from the flow of water over the landscape, we can shape nature to meet our needs. Tile drains, French drains, and simple open-air trenches are all ways to redirect water away from a poorly draining area. Trenches are cheap solutions; simply take a shovel and dig. Though easy for a quick fix, they are not attractive garden features. They also require periodic maintenance to remove debris. Tile drains and French drains are installed subsurface. They cost more money initially, but they last for years or decades, require less maintenance, and are not an eyesore. In my garden, I used French drains to direct water away from my vegetable garden to a naturally low-lying area. I planted the drain outlet full of hardy natives, such as spirea, willow, and red osier dogwood. The resulting rain garden became a low-maintenance but beautiful landscape gem.

When poor drainage is caused by a high water table, there is no easy fix. The best solution is to avoid these areas altogether by watching water on your landscape before you plant or plan your garden. If you're stuck with high water, try building a raised bed to keep plant roots up and out of waterlogged areas. Raised beds, filled with a soil mix, function similarly to container gardens. Because they are removed from the soil ecosystem, they require more attention. They dry out quickly and need more frequent irrigation. You'll need to replace organic matter, fertilizer, and a portion of the soil mix every year in a raised bed.

Size your raised bed for easy access, generally between 3 and 4 feet wide. Select a height anywhere from 6 inches to waist height. Remove existing vegetation. Fork or till the top 6 to 8 inches of native garden soil, and incorporate some of your raised bed soil mix (see box) into the tilled native soil. Build a structure out of lumber, recycled concrete, blocks, or rocks to enclose the bed. Fill the raised bed with soil mix. Raised beds warm quickly, so you'll be able to plant early in the spring.

RECIPE FOR A RAISED BED

When you can't deal with the soil texture you've got, use a raised bed instead.
Use raised beds when . . .

- Soils won't drain
- Gravely soils won't produce
- You don't want to bend over

Soil Mix Recipe

1 part soil
1 part screened compost
1 part porous material (perlite, vermiculite, or coconut coir)
Balanced fertilizer (2 cups per 100 square feet)

CHAPTER 7

CULTIVATION SIMPLIFIED

WHOLE-GARDEN PLANNING FROM THE GROUND UP

Open the door and look out on your garden, or soon-to-be garden. How does it look? How does it feel? Wander around the turf of your lawn and along garden pathways. Are you swimming in microbes? Do your feet feel the spongy resilience of a soil full of organic matter? Is the soil surface covered with an array of living and dead materials? Does your soil feel alive?

These are the questions I ask myself when I'm in the garden. If there's an area that feels sterile and lifeless, or an area overrun by pest and disease, then I need to nurture the living soil to bring it back into balance. To do this I go back to the basics—food, water, shelter, and air. Whether starting a new garden or reinvigorating an existing garden, I use the same perspective. By asking what I can do to improve conditions for the living soil, I find the best tools for the job.

In this chapter, we zoom out to apply what we've learned about growing healthy soils to a whole-garden perspective. We examine our options for cultivating a living soil by integrating biological and mechanical ways to work a soil. In this way, we apply the fundamental soil-building principles to start a new garden, reinvigorate an existing garden, fix problem soils, and perform yearly maintenance tasks.

THE RIGHT TOOL AT THE RIGHT TIME

When I reach into the soil-building toolbox for garden solutions, I find four fundamental tools. The first, and perhaps

For many people, soil conjures images of a freshly tilled field. Tillage, however, is only one of the many tools we can use to cultivate the soil.

most important, tool is planning. I want to match my goals with the natural flow of the garden. I plan where and what to plant based on soil texture, water flow, slopes, and accessibility. The second tool is what I add to the soil. This includes a mulch cover, an incorporated amendment, right water at the right time, or a fertilizer boost. The third tool is what I grow in the soil. Living cover aboveground and living roots belowground add structure, food, and shelter to the living soil. At the same time, the plants that form a living soil cover contribute to a diverse and healthy whole-garden ecosystem.

The fourth and final tool is what I do to cultivate the soil. Most of the time, I let the soil critters do this work for me. Despite all the help from the in-ground gardeners, though, sometimes a little muscle goes a long way. When I need to address a particular problem, speed up the process of improving aeration or drainage, or lay the foundation for a new garden, I go to the actual toolshed. I'll sharpen the tines on my digging fork, the edge of my spade, or even the blades of my rototiller. The right type of cultivation at the right time can put your garden soil on the path the success.

When most people think of cultivating a soil, they first think of tilling—the act of applying mechanical or muscle power to a soil. The early spring rush has old and new gardeners alike lined up to rent or buy the coveted rototiller. In our collective imagination, tillage is almost synonymous with gardening. When imagining a fertile soil, we think of dark brown earth, freshly turned over with a fork or tractor.

There is a good reason that we hold onto this image. Tilling provides many garden benefits. Cutting, slicing, pounding, and grinding soils, we can improve the physical conditions that create good tilth. Deep tillage breaks up compacted layers to improve aeration, drainage, and root growth. Large clods are pounded into smaller, bite-sized aggregates. Shallow tillage creates a smooth seedbed, kills annual weeds, or incorporates fertilizer and amendments. By disturbing the soil, tillage speeds up decomposition and the release of nutrients from organic matter.

For all the above benefits of tillage, though, it is actually the last, not the first, tool for the job of cultivating healthy soils. For starters, it's a lot a work—and not the fun kind. Tilling leads to aching muscles and bent backs. Tractors and rototillers come with noise, gas, dust, and fumes. Tilling at the wrong time or the wrong way can also injure the soil by squeezing soil particles together and destroying a soil's aggregated structure. Tillage also speeds organic matter decomposition, which counters our efforts to build up this vital fraction of the soil.

The most basic reason to use tillage cautiously, however, is its effects on the living creatures in the soil. Fundamentally, tillage disturbs the soil ecosystem. Don't get me wrong; tillage isn't always bad for a living soil. It even stimulates soil life and leads to healthier and more resilient systems. Repeated and intensive tillage over time, however, dramatically simplifies and reduces soil life. The soil starts to rely on your muscle and machines, instead of biology and nature, to create good structure and tilth.

When cultivating soils, it's important to consider all of the tools at our disposal. Instead of tilling or hoeing, I use sheet mulching or organic amendments to eliminate weeds, prepare garden soils, and invite soil organisms to improve the soil. When I focus on the care and feeding of the living soil, it actively turns and churns belowground to improve soil tilth. As the soil food web becomes more and more complex, residues left on the soil surface decompose rapidly, so I don't need to till them under. Mulching diminishes the time spent scratching out weeds. Adding organic matter to the soil surface invites worms to move between layers and improve drainage.

Nonetheless, tillage is a powerful tool. Used cautiously as part of a program to grow a living soil, it can be the right tool at the right time. The art of gardening is choosing why, when, and how to till in order to support, rather than degrade, the living soil.

HOW TO TILL OR NOT TO TILL

Once you've identified the need to cultivate the soil, the next step is to choose how to do it. In principle, I always try to minimize how much I disturb the soil. This means I use the simplest method to get the job done. As the living soil continues to grow, the soil becomes easier to work. Cultivating by hand, even over large areas, becomes less backbreaking and time-consuming. With time, a hand tool or a sheet mulch can replace the tiller to create deep, friable garden beds.

No Till

The least invasive way to till is not to till. A "no-till" approach uses soil organisms, plants, and natural processes to improve the soil. Specific techniques may differ, but the principles remain the same: Grow a healthy soil community by reducing disturbance and building organic matter. Plan your garden around permanent beds that are maintained and developed without tilling. Manage weeds by heavy mulching. Add amendments and residues to the soil surface. Soil organisms, coming to feed on surface residues, incorporate them into the soil. Meanwhile, soils are loosened, lightened, and tilled by burrowing

A no-till approach uses a variety of methods, including permanent beds, sheet mulching, green manures, and the activity of soil organisms, to improve and cultivate soils without tillage.

soil animals and the roots of green manures. If managed properly, soils in a no-till garden are light and soft. Weeds are hand-pulled, and the soil surface is easily massaged into a crumbly and smooth seedbed.

If starting a garden from scratch, the no-till approach uses a thick sheet mulch to kill off competing weeds and lawns. When starting a new garden, this method requires more time and patience than preparing soils with a traditional till. In my experience, it takes about a year for a heavy sheet mulch to decompose and for soil organisms to start building the good tilth for a productive and enjoyable garden. In cool, wet, or clay-rich soils this process can take longer than in light, loamy soil. By keeping the soil community intact, however, a no-till sheet mulch builds a healthy, living soil from the ground up.

Eventually, I like to have most of my garden in permanent, no-till beds. If I use good crop rotations, mulching practices, green manures, and amendments, I find that these are much more productive and have less maintenance requirements than beds that I dig every year. I may, however, jumpstart a bed with traditional tillage and then switch to a more or less no-till approach. I use this hybrid method when I don't have the time or space to wait for no-till sheet mulches to decompose. When dealing with heavy clays, I also like to use a hybrid approach, since a no-till sheet mulch tends to be wet and heavy. I'll lighten the clay with a light tillage the first year and then rely on organic amendments and green manures to continue to improve soil structure in subsequent years.

It's hard to break the tillage habit, but once you see the benefits of a no-till approach, it is hard to go back. Even when I decide my gardens do need a light till, I try to make it as minimal as possible. By using the simple, low-impact methods in the sections that follow, I create a garden that is self resilient and self maintaining.

Sheet Mulching to Start New Gardens

Instead of ripping and tilling lawns and weeds, use sheet mulching to create new garden beds. Sheet mulching reduces the garden workload by advantageously using existing vegetation to build organic matter in place. At the same time, it eliminates the work of tilling, while keeping soil biological communities intact. Although it takes time for mulched residues to decompose, sheet mulching new garden beds results in soils with considerably more organic matter and more active soil biological communities right out of the gate.

Essentially sheet mulching uses the same principle as sheet composting: layering organic material on the soil surface. For starting new garden beds, use a smothering layer of wetted cardboard to kill the turf and weeds. It is possible to plant into the sheet mulch soon after building it by cutting a hole in the cardboard and planting into the original soil. Depending on the tilth and productivity of the native soil, crops may do poorly in the first year, as the soil is in the process of improving. I prefer to build the sheet mulch in the

START A GARDEN THE NO-TILL WAY: SHEET MULCHING STEP-BY-STEP

Collect enough cardboard and green and brown compost material to cover the new garden area. Mow or trim existing vegetation and leave the clippings or trimmings on the soil surface.

Spread cardboard (newspaper also works) over the soil surface to smother living vegetation, ensuring a 4-inch overlap between pieces so there are no gaps for weeds to peek through. Wet the cardboard thoroughly. While it is still wet, begin adding the sheet mulch layers.

For the first layer, spread 1 to 3 inches of nitrogen-rich material, such as manure or grass clippings, evenly over the cardboard. To correct deficiencies or provide nitrogen to the sheet mulch, sprinkle your complete fertilizer over this green layer.

Follow this initial layer with alternating 2- to 6-inch layers of brown, carbon-rich material, such as straw or dried leaves, and green, nitrogen-rich material, such as fresh trimmings or manures. (For a detailed diagram showing how to construct mulch layers, see page 94.) As you add each layer, water it thoroughly to make sure it is moist but not sopping wet. Continue to add layers until you've created a sheet mulch 12 to 18 inches deep. Use a brown mulching material, such as leaves or straw, for the topmost layer to keep the sheet mulch from drying out.

Ideally, let the sheet mulch decompose and mellow in place for six months to a year. To plant immediately, pull back the sheet mulch and cut holes in the cardboard for transplants or large seeds, such as beans and corn. Plant smaller-seeded vegetables and flowers by spreading and planting into a thin layer of compost on top of the sheet mulch. Keep this mulched and well watered as these small seeds tend to dry out.

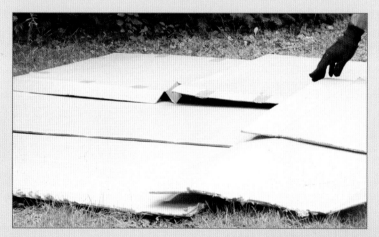

Mow, pull, or trim existing vegetation to ground level. Use a thick layer of wetted cardboard or newspaper to smother existing vegetation.

Start the sheet mulch by covering the wet cardboard with a thin layer of nitrogen-rich material, such as freshly mown grass clippings. Cover this with alternating layers of green and brown material.

Stop adding to the sheet mulch when it is 12 to 18 inches thick. Top with a thick layer of mulch, such as straw or leaves, to conserve moisture.

Plant perennials or transplants right away by pulling back the sheet mulch and cutting a hole into the cardboard. Replace the sheet mulch, leaving a gap around the new plant.

To plant small-seeded annuals, spread a thin layer of compost on top of the sheet mulch. Plant directly into this, but keep it thinly mulched and watered to keep seeds from drying out.

fall, let it decompose over the winter, and plant my annual garden into it the following spring. Planting perennials into a new sheet mulch works out pretty well, particularly if given a starter dose of fertilizer. When the sheet compost and cardboard decompose, a rich garden bed remains behind for the next season.

Work the Soil with Roots

Plant roots in their own right are amazing tilling machines. They expend enormous energy to push through dense soil, loosening it up as they go. When plant roots decompose, they add a sizable amount of organic matter to the soil. As this happens, soils are fluffed and lightened even more, improving structure and drainage.

For this reason, living plant roots are an important part of a no-till system. A bare soil is an opportunity to plant cover crops, whose roots till and prepare the soil for the next garden crop. A forage radish cover crop, planted in the fall and allowed to decompose in the ground after the winterkill, is particularly useful for improving soil structure in a no-till garden.

Let It Rot

Because organic matter is the main ingredient for growing healthy soils, one of the soil grower's major tasks is incorporating organic matter into the soil. Manures and composts are easily added by forking them lightly into the soil. Incorporating plant residues, such as garden debris or cover crops, requires considerably more energy. To reduce this extra effort, no-till gardens simply add organic debris to the soil surface, where it rots with the help of soil microorganisms. Depending on the health and vigor of the soil ecosystem, this can happen very quickly.

To use this technique, either chop or mow cover crops or garden residues onto the soil surface. If very woody, add some nitrogen fertilizer, compost, or manure to speed decomposition. Keep residues well watered. This method has the advantage of acting as a mulch while adding organic matter to the soil. The disadvantage is that some nutrients are lost when they decompose on the soil surface.

Cover It Up

Successful no-till gardening depends on controlling weeds. For this reason, mulching is a key no-till technique. As the mulch rots, it adds organic matter to the garden. In the meantime, it keeps weeds from competing with garden plants. Instead of breaking out the hoe or the tiller, mulch weeds out of existence. Sheet mulching with cardboard in problem spots can accomplish this goal. Continually add organic materials to the soil surface, and keep the soil covered with living plants to strategically keep weeds at bay.

Tilling by Hand

When I do end up digging around in the garden, I like to do it by hand. My digging fork, garden spade, bow rake, and heavy-bladed hoe fulfill most cultivation needs. Not only does this minimize disturbance, it gets me down to the ground level to become intimately familiar with my soils. Aching muscles or a bent fork tine lets me know where to concentrate my efforts to improve soil tilth.

I find that using hand tools complements other no-till techniques. No-till gardening builds soils that are soft and friable. As soils improve in this way, using hand tools becomes less and less work. When a fork slides easily into soft soil, then it does not take long to turn residues in or fluff up a garden bed. When I have optimized soil tilth, I find that one of the only tools I need is a handheld gardening knife. This is my definition of success.

As with any type of tillage, including using mechanized rototillers or tractors, I avoid working the soil when it is too wet or too dry. This is particularly important for the clay soils. Even hand tillage, at the wrong time, can destroy their fragile structure. Not only is this important for maintaining soil structure, soils are much easier to work, particularly by hand, when they have the right moisture content. If soils are too dry, water your garden for a day or two and wait for the soils to become soft and crumbly.

When I do reach for a tool, the tasks I generally perform include mixing amendments and fertilizers into the topsoil with a garden fork. I use a heavy hoe to chop larger weeds at the root or an open-ringed hula hoe to scratch out young weeds between garden plants. I chop residues and cover crops into smaller pieces with a sharp garden spade or hoe. By chopping residues,

I choose hand tools over tillers or tractors any day. A square-bladed spade (not shown), a garden fork, a bow rake, and a heavy-handled hoe (not shown) offer the versatility I need for almost any chopping, digging, scraping, and shaping garden task.

DOUBLE DIGGING STEP-BY-STEP

Double digging, though a lot of work, can jump start a new garden by improving a compacted soil. You will need a flat spade, garden fork, 12-inch board, and a wheelbarrow.

Start by measuring a garden bed 3 to 4 feet wide and 100 feet long. If less than 100 feet, you'll have to supplement with some extra garden soil when you have finished.

1. Stand at one end of the bed. You'll work across the length of the bed from here, in 12-inch wide slices. Use your spade to evenly dig a trench 12 inches wide and 12 inches deep across the width of the bed. As you dig, place the board over the bare soil to distribute your weight and prevent compaction. Place the soil from this first trench in your wheelbarrow for later.

2. Use the fork to loosen the soil at the bottom of the trench, as deeply as the fork will go into the soil.

3. Move the board 12 inches back from its original location. Dig a second trench 12 inches deep by 12 inches wide. Scoop the material from the second trench into the empty space in the first trench. Again loosen the soil at the bottom of the second trench with a garden fork, as deeply as the fork will go.

4. Repeat this process along the entire length of the bed. Continue to scoop the material from the new trench into the empty space of the previous trench. Loosen the subsoil in each trench as you go.

5. When you get to the end of the bed, place the material from the first trench (the soil you placed in the wheelbarrow) into the last trench.

6. Shape the bed by redistributing the soil evenly. Add compost or manure and fork lightly into the top 2 to 3 inches. Rake the seedbed smooth. Now your bed is ready to plant, mulch, and produce. If not planting right away, mulch to protect the bare soil.

I partially incorporate them into the soil. If needed, I use the garden fork to turn them under all the way. I also use a garden spade to chop raw materials into smaller pieces before adding them to the compost pile. Similarly, I break apart heavy clods at the soil surface with the garden spade and fork. To make a new seedbed I use a bow rake to smash surface aggregates and rake them smooth.

It saves me time and effort if I keep my tools in good condition. I frequently sharpen the edges of these tools with a hand file to make them more effective. At the end of the season, I'll clean them and oil them before hanging them up.

I also use hand tools to deal with compacted soils. As mentioned previously, a heavy U-bar tiller can open up deep, compacted layers. Double digging with hand tools is also an extremely effective way to loosen deeper soil layers and improve overall soil aeration and drainage. I don't suggest double digging as a standard practice. Initially, it is extremely time consuming and labor intensive. Because of this, I only use it as an intervention in extreme cases. I think of it as a one-time investment to build a bed that I manage in the future with less intensive and disruptive cultivation techniques.

Essentially, double digging removes the topsoil in measured trenches. This allows you to fork and loosen the subsoil, which in effect raises the level of the entire bed. The topsoil is replaced as you go and enriched with organic matter and other amendments. If done correctly, you'll end up with a bed 2 to 3 inches higher than when you started.

Mechanized Cultivation

Although I prefer to keep things simple when possible, I do find occasion to use machines. There is rarely cause to use mechanical tillers in small-scale home gardens, which benefit from the low-impact use of hand tools. I only consider mechanical cultivators for larger-scale gardens or small farms, when they can help me reach my soil-growing goals more efficiently. Even then, I keep these intensive measures in reserve, when more extreme situations call for a strategic use of force.

In my opinion, it is best to minimize how often I do till. I might use a heavy-handed, mechanical tilling method on a one-time, as-needed basis, but I won't make it a part of my regular garden care routine. If I decide to till a soil, I'll try to accomplish several cultivation tasks at once. For instance, if tilling to improve drainage, I'll use the opportunity to incorporate organic amendments during the same pass.

At the scale of a small farm, tractors are sometimes useful to cover ground on a regular basis. Even at this larger scale, using methods that reduce the amount and intensity of tillage creates healthier and more productive soil ecosystems. Examples of this include using cover crops to break up the subsoil and letting residues rot on the soil surface instead of turning them under with the tiller. If possible, avoid using tractors and rototillers for routine tasks such as fertilizer incorporation or seedbed preparation. If incorporating residues or loosening a soil, use the minimum number of passes needed to accomplish your goals. If necessary, break out the rake and the fork after using the tiller to put the finishing touches on a garden bed.

Rototillers and tractors apply more force and weight than hand tools, which leads to a greater risk of compacting your soil. In this case, it is even more important not to till overly wet or dry soils. Before you till, test whether the soil has the right moisture content. A scoop of soil should feel like a scoop of well-watered compost, slightly moist but without excess water. Even in soils with the right moisture content, repeated tillage with mechanized implements can lead to a compacted layer known as plow pan. This happens because the tiller tines are repeatedly applied at the same depth.

Generally, I consider using mechanized tillers for two purposes: (1) preparing the soil for a new garden or (2) incorporating green manures and cover crops. Small rotary tillers are good for both of these purposes. To minimize the number of passes needed, I first

Mechanical tillage effectively incorporates a mature cover crop into the soil. Use the minimum number of passes needed to turn the residue under, then let the soil microbes finish the job of decomposition.

mow or chop existing vegetation or cover crops to the soil surface. I follow this with one or two rototiller passes. I water well and give the residues a few days to a week to break down. I'll follow this up with another pass or two to fully incorporate the material. At this point, I ideally let the material decompose for several weeks before finishing the bed preparation with hand tools. I find that this staggered technique works better than trying to fully incorporate fresh material. Doing this gives the microbes time to help me prepare the new garden bed.

For larger areas, judicious use of disc harrows attached to a tractor can help to loosen soils, incorporate cover crops, mix in amendments, or smooth a seedbed. Once again, these implements are best used only as occasional tools. Focus on adding and maintaining organic material in the soil, rather than mechanized tillage as the primary cultivation tool.

STARTING A NEW GARDEN

Every new garden is unique. When starting from scratch, I return to the soil to ask how I can create the optimum conditions to grow a living soil. Depending on the soil, sometimes I just need to maintain the existing living, healthy soil system. Sometimes, I need to recreate it. Taking the time to observe and plan helps put the new garden on the path to success.

Whole-Garden Planning

The first step in whole-garden planning is observation. Ideally, take a season or a year to watch the flow of water, energy, and gravity across the landscape. Use the simple jar test and soil-quality indicators (see Chapter 1) for different areas of your garden to determine soil texture and the status of soil health. Get a laboratory soil test to further identify nutrient deficiencies you need to address. Working with my natural garden system—i.e., going with the flow—establishes a foundation for further growth.

Now take some time to think about your gardening and soil-building goals. What types of gardening areas do you want to establish? For what purpose? How often will you visit them? What areas do you want to see every day? To build soil, do you plan on making your own compost? Do you want to dedicate areas of your garden for growing compost materials? Is there space available to stockpile collected compost materials? Will you include extra space for green manure and cover crop rotations?

Garden Zones and Sectors

Zones and sectors are a fundamental permaculture design principle. Zones form the basis of garden planning based on how often you need to use or take care of an area. With this model, areas that you visit frequently, such as a bed of salad greens, a worm bin, or a floral display garden, are located close to the home. Areas that need less attention or that provide only periodic harvests, such as fruit trees, compost bins, or long-season vegetables, are located farther from the home. This type of planning makes the best use of energy across the landscape. Think logically about how often and what paths you will take to move across the garden in order to maximize the efficiency of your time, energy, and enjoyment.

In addition to zones, you'll also want to think about sectors. Sectors divide the garden based on its inherent limitations and possibilities. You'll use this information to modify how you situate your garden zones across your whole garden plan. After observing your garden, map separate sectors based on the following natural attributes:

1. **Soil properties.** Are there observable differences in the feel of your soil across the garden? Did your soil-test report determine measurable differences in texture or other soil properties? If so, delineate areas that are more clay or sand rich. You may have areas where topsoil was damaged or removed. If so, map these as well. Topsoil-poor sectors will require more attention and soil-building work.

2. **Drainage and flooding.** After watching water across the landscape, map areas that tend to collect water and areas that tend to discharge water. Specifically note areas where water ponds or pools or areas prone to flooding in a heavy rain.

3. **Sunlight.** The availability of both winter and summer sun is extremely important in garden planning. Map areas that receive maximum winter and summer sun, and plant your garden accordingly. Use trees and shrubs strategically to provide shady microclimates, without unnecessarily removing precious sunlight.

4. **Winds and views.** Are there sectors of the garden where you want to block outside inputs? Are there edges of your garden that are subject to heavy winds or ugly views? These edges are good locations for tall borders that can block or dampen these effects.

The presence and direction of slopes is a final consideration in garden planning. Depending on how we treat them, garden slopes can be assets or liabilities. Slopes can help to collect and direct sunlight and water, or they can lead to reduced soil quality by the action of water washing away precious topsoil. Identify any slopes in your garden, even gentle ones. In your garden planning, consider how slope aspect (direction) affects both winter and summer sun. Think about whether you can build trenches or swales diagonally across a slope to capture and slow water. Are there possibilities at the base of the slope for water to collect in a rain garden or a pond? On more severe slopes, you may consider planting brush or tree breaks across a slope to prevent and capture eroding soil.

Because erosion is a concern on sloping land, it becomes even more important to minimize disturbance. Always build beds on contour, horizontal to the direction of the slope. Consider building terraces, and use permanent beds to minimize tillage. Cover any bare areas with thick mulches, and establish groundcovers that conserve soil in place. On sloping land, keeping topsoil in place is the foundation for growing healthy, living soils.

Now that you understand your natural garden system, it's time to match goals with garden capabilities. I like to start by making a physical map of imaginary zones and sectors in my garden. I call them imaginary because they don't have hard and fast boundaries. Remember to keep these areas fluid and blended so that they can change according to your garden needs and observations. Include soil texture and information from soil test results on these maps. In particular, delineate areas that have low or high pH or excessive salt content.

Building beds along the contour of a slope, as shown in these ancient rice terraces, conserves and captures soil. Use this same principle even for the gentlest slopes in the backyard garden.

Next, make a list of all the things you want to grow or garden spaces that you want to include. As a soil grower, this is the also time to consider your soil-building plan. To grow great soils, you'll need to include space for crop rotations and green manures. You may want to rotate areas in and out of production as you apply heavy sheet mulches. Make sure the garden includes the extra space you'll need to grow soil. When considering garden layout, choose the best locations to make crop rotations easy and practical.

Every year you'll need to amend the soil with organic material. What can you access on your property, in your neighborhood, or from your city? What amount of compost will you need annually? If making compost at home, do you have enough space for piles and stockpiled materials? If importing manures and other raw materials, do you have a good place to stage, store, and unload transported material? Among your perennial shrubs, hedges, and trees are there species that will provide great mulching, chipping, or compost material? Is it easy to access these? Have you considered planting nitrogen-fixing perennials that you can chip to add fertility to the garden? Answering these questions should help you flush out your whole-garden plan, while including essential elements needed to grow great soil.

Now, take your map of garden attributes and your wish list of garden goals. First, use sectors to identify areas that may limit garden uses. If you've identified areas that are poorly drained, are high in salts, or have high or low pH, then consult garden guides to find which plants will tolerate these conditions. Alternatively, set these areas aside for storage, composts, or other soil-building activities. Choose the best plants to match each sector's advantages and disadvantages. Next, think about capturing sunlight and water. How can you structure the vertical height and horizontal contour of your garden to capture both while reducing wind and maintaining nice views? Finally, think about your zones, based on frequency of use. Among the plants for features that you've identified for each sector, place the one that you visit the most closest to your garden door. Whether you break out the art supplies to create the master garden plan or simply scribble notes as you think through these questions, having a plan in hand makes the best use of what you have while building the foundation for growing great soils into the future.

Preparing the Soil

Now that you have your plan in hand, it's time to prepare the soil for planting. How you do this depends on what you are planting and the condition of the soil. Often, soils abandoned for many years are extremely fertile, healthy soils. This is particularly true where weeds, grasses, and existing plants grow into an untended jungle that covers the soil and adds organic matter, naturally. These naturally productive soils often have loamy textures that need little extra help. In these cases, I like to use no-till methods to mulch out existing vegetation without disturbing the natural good tilth of the soil.

In other cases, I will often till for the initial soil preparation. This is useful for compacted clays, areas that have lost topsoil, or nutrient-deficient sandy soils. I'll loosen up deeper soil layers and incorporate organic material and fertilizers throughout the 6- to 8-inch rooting depth. This helps correct nutrient imbalances and drainage issues. By initially improving soils in this way, I get increased plant productivity right from the start. Increased productivity, if returned to the garden soil, results in increased organic matter. This, in turn, begins a self-reinforcing cycle to continually improve and grow living soils.

Regardless of whether I am tilling or using no-till methods to start a new garden, lawn, or landscape, it is important to start preparing the soil well in advance of planting. With soil-test results in hand, I can incorporate the right amendments and fertilizers as I till. If added early enough, fertilizers, lime materials, and amendments have time to decompose and activate in the soil. The following sections outline the step-by-step way to prepare soils for new lawns, vegetable gardens, or perennials.

Establishing a Lawn

Correct any soil drainage problems before establishing a lawn. If necessary, install French drains or tile drains to direct water away from lawn areas. Lawns do well on gently sloping areas that allow water to drain naturally. Incorporate 1 to 2 inches of compost or ½ to 1 inch of composted manure to a 6- to 8-inch depth. For heavy clay soils, incorporate another 1 inch of compost or another organic amendment.

Sow cool-season grasses in the fall, at least four weeks before the first hard frost. In warmer climates, use warm-season grasses, sown in the spring to early summer. Consult your local extension service or garden center for species that do best in your area and for your soil type. Ideally, prepare the soil about a month in advance of seeding. Use your soil test results to determine fertilization needs. Till under any existing vegetation and give it a week or two to decompose. Add the required amount of lime, fertilizer, and compost to the soil surface and till to a depth of 6 inches. Rake the soil surface smooth. Broadcast the grass seed and tamp it into the soil for good contact. For larger areas, rent a lawn roller. For smaller areas, use the top of your rake or spade head. Mulch the newly sown area thickly and be sure to keep it moist in the early stages of growth. Allow the lawn to grow for about a month before cautiously mowing or walking over it. If there are areas where seed did not establish well, repeat the sowing, watering, and mulching in spots.

Establishing a Vegetable Garden

For vegetable gardens, decide whether you can use till or no-till methods to establish garden beds. Submit soil tests to laboratories in the early fall. If using a no-till approach, sheet mulch new garden beds at this time as well. Include enough manure and nitrogen-rich green materials in the sheet mulch layers to partially meet the garden's nitrogen needs. Sprinkle the recommended lime application between the compost layers. Keep the sheet mulch well mulched and moist throughout the winter. In the early spring, spread your remaining fertilizers and gently fork them into the sheet mulch. If the sheet mulch is only partially decomposed, pull it back in pockets to transplant seedlings into mineral soils. For more decomposed sheet mulches, plant directly into the new topsoil. If seeding, spread a thin layer of compost and seed directly into the compost.

If starting a new garden by tillage, turn under existing vegetation in the early fall. This is also a good time to incorporate lime and manures, if needed. Allow the residues to decompose for a week or two. Rake the seedbed smooth and plant a winter cover crop. In the early spring, turn in the cover crop residue by hand or with a rotary tiller. Incorporate fertilizers at the recommended rate. Wait one to two weeks for fertilizers and residues to decompose. Shape and rake smooth the new beds to plant spring crops. If not planting right away, keep the beds well mulched. Consider planting a quick-growing buckwheat to cover spring soils before it is time to seed summer vegetables.

Establishing a Perennial Garden

Because perennial gardens are a more or less permanent feature, correct any drainage problems before establishment. If necessary, double dig or deep rip soils to remedy compaction. If poor drainage is not an issue, no-till methods are a good way to start a perennial garden, as long as individual plants are well supplied with the nutrients they need. If you have the material at hand, cover the entire garden area in a thick sheet mulch. You can plant directly into this sheet mulch by cutting a hole in the bottom cardboard or give the entire area a year or so to compost and mellow. If planting directly into sheet mulch, be careful to keep mulch pulled back from the base of perennials in order to prevent rot. Because this leaves the base of plants open for weeds that have not been smothered by the sheet mulch, be prepared to hand pull around new plantings. If tilling a new perennial garden, use more or less the same process described for establishing a new lawn. Turn under existing vegetation. Incorporate lime, fertilizer, and 1 to 2 inches of finished compost or ½ to 1 inch of composted manure. Plant perennials soon after bed preparation and thoroughly mulch the entire area with 3 to 4 inches of organic material.

Most perennials are best planted in the early fall, to give their roots a head start on establishing before the spring rush. To plant, dig a hole that is wider than the transplanted stock. Loosen the side of the hole with a garden fork and mix the recommended phosphorus and lime thoroughly in the loosened soil. Backfill the hole about three-quarters full with the removed soil. To remove any air pockets near the roots, saturate the hole with water and wiggle the new plant into the slurry. Continue backfilling with soil. Replace the mulch, leaving space for air circulation around the base of the new tree or shrub. Because nitrogen and potassium will stimulate new growth, topdress with these nutrients in the later winter or early spring to prepare plants for the growing season.

REBUILD YOUR GARDEN FROM THE GROUND UP

Many times people want to know how to radically rebuild a garden or lawn. When unproductive year after year, it's natural to want to rip it up and start over. Who doesn't want a clean slate? The impulse to rebuild from scratch, however, sets your garden way back. It's far better to use what you already have to build better soils from the ground up.

It's true that not all soils are created equal. It's particularly difficult when gardens are bereft of topsoil because of construction, erosion, or natural conditions. Trucking in better soil from somewhere else seems like the obvious solution, but although this can work for raised beds or container gardens, it doesn't work well at the garden scale. For starters, it is expensive. Secondly, because transported soil is often sterile, you'll have to start from scratch with soil-growing techniques to build the living soil community.

Planting perennial trees and shrubs in the fall gives roots time to establish before the spring flush. Plant directly into the new sheet mulch or newly tilled perennial bed. Mulch well, but wait to add nitrogen fertilizer until the late winter or early spring.

Trucking in different soil also often leads to drainage problems. Finally, new surface soil is simply a thin bandage that just covers underlying problems of deeper layers. In most cases, we will have to address these sooner or later.

So what can we do to reinvigorate suffering garden soils? First, take a step back. Do your plants match the capabilities of your soil? Look for obvious mismatches, like drought-loving plants in a poorly draining soil or acid-intolerant species under low-pH conditions. Instead of ripping up your soil, take out the ill-suited plants and replace them with more appropriate varieties. Overall garden productivity, and gardener satisfaction, will definitely increase.

Examine obvious physical limitations of your soil. If poor drainage or hardpan is a problem, apply solutions from Chapter 6. If waterlogged conditions can't be avoided, consider planting wetland plants or digging a pond to adapt to the soil's natural condition. For lawns, compaction is a regular problem. Use the tines of your garden fork or a lawn core cultivator to periodically aerate the soil.

Next, look at soil test results. Excessive salts, a pH less than 5 or greater than 8, and major nutrient deficiencies are red flags that must be addressed. Salt buildup requires

Help compacted lawns breathe by periodically aerating the soil with a lawn aerator, core cultivator, or the tines of your garden fork.

extensive remediation and site-specific consultation. Imbalanced pH and nutrient deficiencies often go hand in hand. Address these imbalances by applying corrective lime and fertilizer. To affect the entire rooting zone, till these materials into an annual garden. For perennial gardens and lawns, spread evenly over the grass or under the perennial mulch. Scratch these materials lightly into the soil surface and water. Continue to get soil tests and correct any imbalances on an annual basis, until soils improve.

Sometimes, soil texture is a major limitation. For heavy clay soils, use soil conditioners to break up dense clods. Gypsum, fresh or composted leaves, and even low-quality sawdust work wonders when incorporated for several years in a row. If you have the time, plant a perennial cover crop, such as alfalfa, to prepare the soil for several years before replanting.

After addressing any obvious chemical or physical limitations, the next step in garden renewal is to rebuild your soil from the ground up. Instead of wasting money and time on imported and lifeless soil, use materials at hand to build soil on the existing soil surface. Build your own topsoil by continually adding organic amendments, sheet composts, and mulches. Gradually, this feeds and stimulates the soil microbes to improve belowground conditions. Use the raw power of composting in place to restore the soil community. Apply thick sheet mulches around well-established perennials. An aggressive trench compost rotation can quickly rebuild the soil community belowground in an annual garden. Using organic matter in this way mimics natural soil processes to improve and rebuild suffering soils for the long-term.

A SOIL GROWER'S CALENDAR

Planning is the key to any successful garden. When we focus on the soil, our annual chores revolve around procuring, storing, and preparing organic material. Whether building compost, planting cover crops, raking leaves, or adding fertilizers, we need to think ahead. To prepare for the next spring, a soil grower's calendar starts in the late summer.

Over the years, my garden has adapted to building soil as it grows. I've planted perennials that satisfy many of my organic matter needs. I chip and shred trimmings as I prune over the winter. I know when and where to procure neighborhood grass clippings and tree leaves. Many of my beds are self-mulching and require little assistance from me. My primary job is to observe my garden's changes through the seasons and to respond creatively with the tools I have at hand. When my garden is covered in a living and diverse jungle, I am confident that it feeds and is fed by a living and diverse jungle of life belowground.

SOIL THROUGH THE SEASONS: A SOIL-GROWER'S YEARLY CALENDAR

Late Summer/Early Fall

Submit soil tests.

Plant cover crops.

Turn compost.

Put beds to sleep—chop residues into soils and mulch the surface for winter.

Spread and incorporate manures, fresh leaves, and yard debris.

Sheet mulch new garden areas.

Till and lime new garden areas. Add organic phosphorus fertilizers.

Plant perennials.

Plant cool-season lawns.

Spring

Add fertilizers and amendments to soils.

Incorporate winterkilled fall and winter cover crops.

Incorporate compost and manures.

Prepare new seedbeds.

Remove mulch on heavy soils to let them warm.

Plant spring crops.

Plant quick-growing cover crops in summer vegetables beds.

Build and turn compost.

Late Fall/Winter

In cold climates, move the worm bin indoors.

Plan for soil-building needs.

Complete the Whole-Soil Fertility Worksheet to determine fertilizer needs.

Plan crop rotations.

Stockpile compost materials: manures, cardboard, leaves, grass clippings, etc.

Build leaf mold compost piles.

Build manure compost piles.

Kill cover crops.

Shred leaves for mulch.

Prune trees and shrubs. Chip trimmings for mulch.

Plant perennials.

Fertilize perennials in the late winter.

Summer

Build and turn compost.

Keep garden and compost well-watered.

Plant warm-season cover crops.

Plant warm-season grasses.

Implement crop rotations and intercrop plantings.

Pull up early spring cover crops. Use as mulch.

Pull weeds. Use non-noxious weeds to mulch soil surface.

Mulch summer vegetables.

Side-dress heavy feeders with nitrogen-rich material.

Use foliar spray fertilizers as needed.

Collect grass clippings from neighbors.

ADDITIONAL RESOURCES

SOIL PROPERTIES

Estimating Soil Moisture by Feel and Appearance. US Department of Agriculture, Natural Resources Conservation Service (USDA-NRCS). 1998. www.nrcs.usda.gov/Internet/FSE_DOCUMENTS/nrcs144p2_051845.pdf

National Soils. USDA-NRCS. www.nrcs.usda.gov/wps/portal/nrcs/site/soils/home

Web Soil Survey. USDA-NRCS. websoilsurvey.sc.egov.usda.gov

SOIL HEALTH

Gershuny, Grace. *Start with the Soil.* Emmaus, PA: Rodale Press, 1997.

Gugino, B. K., O. J. Idowu, et al. *Cornell Soil Health Assessment Training Manual,* 2nd ed. Geneva, NY: Cornell University, 2009.

Lowenfels, Jeff, and Wayne Lewis. *Teaming with Microbes: The Organic Gardener's Guide to the Soil Food Web,* revised ed. Portland, OR: Timber Press, 2010.

Magdoff, Fred, and Harold van Es. *Building Soil for Better Crops,* 3rd ed. Beltsville, MD: SARE, 2010.

Tugel, A. J., A. M. Lewandowski, and D. Happe-von Arb, eds. Ankeny *Soil Biology Primer.* IA: Soil and Water Conservation Society, 2000.

Soil Quality Test Kit Guide and Guide. USDA-NRCS, Soil Quality Institute, 2001.

Willamette Valley Soil Quality Card Guide (EM-8710-E). Corvallis, OR: Oregon State University, 2009.

PERMACULTURE

Fukuoka, Matsunoba. *One Straw Revolution.* Emmaus, PA: Rodale Press, 1978.

Hart, Robert. *Forest Gardening: Cultivating and Edible Landscape,* 2nd ed. Chelsea, VT: Chelsea Green Publications, 1996.

Kourik, Robert. *Designing and Maintaining Your Edible Landscape Naturally.* Santa Rosa, CA: Metamorphic Press, 1986.

Mollison, Bill. *Introduction to Permaculture.* 2nd ed. Tyalgum, Australia: Tagari Publications, 1999.

Shein, Christopher. *The Vegetable Gardener's Guide to Permaculture: Creating an Edible Ecosystem.* Portland, OR: Timber Press, 2013.

COMPOST AND AMENDMENTS

Applehof, Mary. *Worms Eat My Garbage: How to Set Up and Maintain a Worm Composting System.* Kalamazoo, MI: Flower Press, 1982.

Jenkins, Joseph. *The Humanure Handbook,* 3rd ed. Grove City, PA: Joseph Jenkins, Inc., 2005.

Martin, Deborah L. and Grace Gershuny. *The Rodale Book of Composting: Easy Methods for Every Gardener.* Emmaus, PA: Rodale Books, 1992.

COVER CROPS

Björkman, Thomas. *Cover Crops for Vegetable Growers*. Cornell University. covercrops.cals
.cornell.edu

Clark, Andy, ed. *Managing Cover Crops Profitably*, 3rd ed. Beltsville, MD: Sustainable Agriculture
Research & Education (SARE), 2007.

Hess, Anna. *Homegrown Humus: Cover Crops in a No-till Garden*. Wetknee Books, 2013.

NO-TILL GARDENING

Lanza, Patricia. *Lasagna Gardening*. Emmaus, PA: Rodale Press, 1998.

Stout, Ruth. *Gardening Without Work*. NY: Devin-Adair, 1963.

FERTILIZERS

California Plant Health Association. *Western Fertilizer Handbook*, 9th ed. Long Grove, IL:
Waveland Press, Inc., 2002.

Andrews, N., D. Sullivan, K. Pool, and J. Julian. *Organic Fertilizer and Cover Crop Calculator*.
Oregon State University Extension Small Farms Program. smallfarms.oregonstate.edu/
calculator

Sullivan, Dan. *Estimating Plant-available Nitrogen from Manure*. Corvallis, OR: Oregon State
University Extension Service, 2008.

SUSTAINABLE DESIGN

ATTRA—The National Sustainable Agriculture Information Service. National Center for
Appropriate Technology (NCAT). attra.ncat.org
(a wealth of downloadable material, including designs and blueprints for sustainable farm
and garden practices)

Greywater Action: For a Sustainable Water Culture. www.greywateraction.org
(information and home-scale designs for sustainable water use)

Sustainable Agriculture, Research, & Education (SARE) Learning Center, National
Institute of Food and Agriculture, US Department of Agriculture, 2012. www.sare.org/
learning-center

VEGETABLE PRODUCTION

Bradley, Fern Marshall, ed. *Rodale's All-New Encyclopedia of Organic Gardening: The Indispensable
Resource for Every Gardener*. Emmaus, PA: Rodale Press, 1992.

Coleman, Eliot. *The New Organic Grower*. Chelsea, VT: Chelsea Green Publications, 1989.

Jeavons, John. *How to Grow More Vegetables*. 9th ed. Berkeley, CA: Ten Speed Press, 2006.

Mohler, Charles L. and Sue Ellen Johnson, eds. *Crop Rotations on Organic Farms: A Planning
Manual*. Ithaca, NY: Natural Resources, Agriculture and Economics Service (NRAES), 2009.

Oregon Vegetables: Vegetable Production Guides. Oregon State University, Department of
Horticulture.

INDEX

ABOUT THE AUTHOR

Elizabeth (Ea) Murphy is a soil scientist, writer, farm enthusiast, and gardener. Her love of soils and food takes her on adventures in her own backyard and around the world. For more soil stories, advice, and discoveries follow her blog at www.dirtylittlesecrets.rocks.

IMAGE CREDITS

All photographs by Crystal Liepa Photography and illustrations by Bill Kersey except as noted below.

Cool Springs Press: front cover; 9; 62; 89; 108; 117; 163; 189

iStock: 14, water (GusGus); 74 (Artfoliophoto); 105 (mtreasure); 112 (Willowpix); 116 (susandaniels); 118 (Aaron McCoy); 155, top (PICSUNV)

Shutterstock: 12 (Ed Phillips); 13 (Scott Chan); 14, air (Nemeziya), mineral (Sergey Nivens), center right (daseaford), bottom right (Stephen Rudolph);18 (Just Keep Drawing); 21, left (udra11), right (Denis and Yulia Pogostins); 23 (mashe); 24, top left (Jubal Harshaw); 26 (pornvit_v); 33 (Henrik Larsson); 34 (exopixel); 36, top (R Weir Works); 37 (Stefano Lunardi); 38 (BMJ); 39, center (isak55); 44, top* (Paul D Smith), bottom (w.g.design); 45 (StudioZ); 46 (Steve Martin); 53 (yuris); 56 (Emilio100); 57, top right (Laitr Keiows), bottom (Vahan Abrahamyan); 68 (Oleg Kirillov); 69 (Alena Brozova); 70, top (J. Bicking), bottom (Mikosot); 76, top (Allen W Yoo), bottom (Aigars Reinholds); 78 (Meryll); 79, right* (Bertold Werkmann); 83 (Grigorev Mikhail); 84 (Dani Vincek); 85 (Nobuhiro Asada); 86 (Olaf Schulz); 90 (Humannet); 93 (Candus Camera); 98 (Lane V. Erickson); 103 (Christine Norton); 104, top (audaxl), bottom (Alison Hancock); 126 (1000 Words); 131 (svrid79); 136 (Le Do); 137 (Olaf Speier); 141 (Catalin Petolea); 144, top (sanddebeautheil), bottom (Samuel Acosta); 145 (bikeriderlondon); 147 (Frank11); 152 (Jeanne McRight); 153 (wsf-s); 154 (Irina Fischer); 159, left (Julija Sapic), right (Kraiwut); 165 (basel101658); 167 (Alison Hancock); 168 (Dusan Zidar); 170 (sit); 181 (Toa55); 184 (Cristal Tran)

USDA Natural Resources Conservation Service: all microscopic images (24 except top left, 54, 64, and in illustrations)

Wikimedia Commons: 50, right (Agronom/CC-BY-SA-3.0)

* Also used in illustration